AF300939

Design and Analysis of Combinatorially Assembled Optical Systems

Von der Fakultät für Maschinenwesen der Rheinisch-Westfälischen Technischen Hochschule Aachen zur Erlangung des akademischen Grades eines Doktors der Ingenieurwissenschaften genehmigte Dissertation

vorgelegt von

Max Christian Funck

Berichter: Universitätsprofessor Dr. rer. nat. Peter Loosen
Universitätsprofessor Dr.-Ing. Robert Heinrich Schmitt

Tag der mündlichen Prüfung: 22. Dezember 2011

Bibliografische Information der Deutschen Nationalbibliothek
Die Deutsche Nationalbibliothek verzeichnet diese Publikation in der
Deutschen Nationalbibliografie; detaillierte bibliografische Daten sind
im Internet über http://dnb.d-nb.de abrufbar.
1. Aufl. - Göttingen : Cuvillier, 2012
   Zugl.: (RWTH) Aachen, Univ., Diss., 2011

   978-3-86955-990-2

D 82 (Diss. RWTH Aachen University, 2011)

© CUVILLIER VERLAG, Göttingen 2012
   Nonnenstieg 8, 37075 Göttingen
   Telefon: 0551-54724-0
   Telefax: 0551-54724-21
   www.cuvillier.de

   978-3-86955-990-2

Abstract

Producing high quality optical systems at reasonable cost is a challenge and solutions enabling cost-efficient production in small to medium quantities need to be developed in order for Germany to stay competitive on the global market. Both, performance and cost are largely influenced by manufacturing and assembly tolerances and frequently compensation such as alignment is required to achieve the demanded performance.

This dissertation systematically develops combinatorial assembly as a compensation strategy which does not require iterations and as such is suitable for automated production of small series. A pool of components and subassemblies, necessary for the assembly of a series of systems, is characterized prior to the assembly and measurement results are stored in a database. Then, optimal component combinations are found during a model-based selection process.

The application of combinatorial assembly to optical systems requires a delicate choice of parameters for characterization, modules and tolerances before components are manufactured. Predicting the as-built performance of combinatorially assembled systems with high accuracy is therefore necessary and a dedicated tolerance analysis concept based on Monte Carlo analyses is proposed. The concept is universally applicable to problems that can be modelled with ray-tracing and is implemented using a combination of ray-tracing software, logic calculator and database. This makes it possible to accurately analyze the impact of tolerances, production volume and additional uncertainties on the performance of combinatorially assembled optical systems.

For optimal compensation, tolerance distributions should match each other in a specific way and it is illustrated that this can be difficult to realize for some lens designs due to manufacturing limits. In order to reduce this restriction, design strategies increasing combinatorial compensation are derived. Adapting the optical design from the outset to suit combinatorial assembly can shift tolerance sensitivities from one component to another. Compensation can be enhanced and the influence of uncharacterized parameters reduced. In using combinatorial assembly in conjunction with desensitization, systems with higher nominal performance yet reduced tolerance demands can be build. This is an entirely new approach and a first step towards a more integrated development of optical systems.

Zusammenfassung

Optische Systeme hoher Qualität zu vertretbaren Kosten zu produzieren, ist eine große Herausforderung. Damit Deutschland international wettbewerbsfähig bleibt, ist vor allem eine kosteneffiziente Produktion kleiner und mittlerer Serien erforderlich. Sowohl die optische Güte als auch die Produktionskosten werden dabei maßgeblich von Fertigungs- und Montagetoleranzen bestimmt und häufig sind Kompensationsmethoden wie die Justage erforderlich, um hohe Leistungsanforderungen erreichen zu können.

Diese Dissertation entwickelt mit der kombinatorischen Montage eine Kompensationsmethode, die nicht auf iterative Prozesse angewiesen ist und sich so für die automatisierte Kleinserienmontage eignet. Dabei werden die für eine Serie erforderlichen Komponenten und Baugruppen vor der Montage charakterisiert, die Messergebnisse in einer Datenbank erfasst und im Anschluss daran optimale Bauteilkombinationen modellbasiert ausgewählt.

Die Anwendung der kombinatorischen Montage auf optische Systeme erfordert im Vorfeld der Fertigung eine gezielte Auswahl der zu charakterisierenden Größen, Baugruppen und Toleranzen durch den Optikentwickler. Es ist darum erforderlich, die Güte kombinatorisch montierter Systeme präzise vorherzuberechnen und ein entsprechendes Toleranzanalyse-Konzept basierend auf Monte Carlo Analysen wird vorgestellt. Das Konzept ist universell auf Optiken anwendbar, die mittels Ray-Tracing modelliert werden können und als eine Kombination von Ray-Tracing Software, Logikberechnung und Datenbank implementiert. Dadurch ist es möglich, den Einfluss von Toleranzen, Produktionsumfang und zusätzlichen Unsicherheiten auf die Eigenschaften kombinatorisch montierter optischer Systeme exakt zu berechnen.

Um eine optimale Kompensation zu erreichen, müssen die Toleranzverteilungen der Bauteile und Baugruppen in besonderer Weise zueinander passen. Dies ist aufgrund von Fertigungsgrenzen nicht für jede Optik zu erreichen und schränkt die Anwendbarkeit der Methode ein. Durch Anwendung von Design-Strategien, die das Optik-Design von vorneherein auf eine Anwendung der kombinatorischen Montage ausrichten, können jedoch die Kompensation verstärkt und der Einfluss nicht charakterisierter Parameter verringert werden. Die Verwendung der kombinatorischen Montage im Zusammenhang mit Desensitivierung erlaubt es damit, Systeme mit höherer Nominalleistung bei reduzierten Toleranz-forderungen umzusetzen. Dies stellt einen völlig neuen Ansatz und einen ersten Schritt hin zu einer stärker integrierten Entwicklung optischer Systeme dar.

Acknowledgements

This thesis is based on work carried out during my time as a research scientist at the Chair for Technology of Optical Systems at RWTH Aachen University and at the Fraunhofer Institute for Laser Technology (ILT).

Many people have contributed to this work in many different ways:

I am very thankful to Prof. Dr. Peter Loosen, Head of the Chair for Technology of Optical Systems and Vice Director of the Fraunhofer ILT for his supervision, advice and critical review of the thesis. I would also like to thank my second advisor Prof. Dr. Robert Schmitt, Head of the Chair of Metrology and Quality Management and Chairman Prof. Dr. Feldhusen, Head of the Institute for Engineering Design.

Martin Traub, Jörg Diettrich as well as Prof. Leo Beckmann were the origin for many valuable and inspiring discussions on lens design which I thoroughly enjoyed. I thank them and Jochen Stollenwerk for their comments and remarks with respect to this work.

Special thanks deserve all past and present members of the optics group at RWTH Aachen, the numerous students who assisted me as well as many colleagues from Fraunhofer ILT that made my time there an enjoyable experience. In particular, I wish to thank Jan Dolkemeyer and Valentin Morasch for the fabulous time we had and for their friendship.

Kerstin, I am incredibly thankful for your patience, trust and love. Without you, this thesis might have never come to an end.

Finally, my deepest gratitude goes to my parents for their endless and unconditional support throughout my life and academic career as well as to both of my sisters.

Aachen, December 2011   Max Funck

| Content

1   Introduction

Optical systems and instruments are synonymous for high precision, performance and quality and the requirements to guide light to a sharp focus are indeed very demanding. In terms of geometrical optics, the distance any two rays travel from an object point to an image point must agree within a quarter of the wavelength of light*. In the visible 100 nm is the limit, while for extreme ultra violet light, used in the latest semiconductor technology, the limit is reduced to a few nanometers.

Like only few other countries Germany has a long and successful history in developing and producing high performance optical systems. This focus has remained till today and companies are concentrating on small to medium quantities of specialized and customized products. The specialization is made possible by a large number of skilled designers and engineers but also due to competition from countries with low labor cost. Of all the produced optical systems the largest share in sales volumes is reported for innovative applications in production technology, metrology and life science [OPTE07].

The demand for precise optical instruments is expected to rise even further. This is supported by the observation that optical devices are shrinking in size, that wavelengths and light pulses are getting shorter and that the amount of information transmitted by optical systems is steadily growing [HERI06]. At the same time, however, and as optical systems are increasingly part of consumer products, drastic price and time demands develop. It has long been identified that solutions enabling cost-efficient production of high performance optical systems in small to medium quantities need to be developed in order to stay competitive on the global market [SIEG02].

Three factors with a large impact on performance and cost of optical systems can be identified: optical design, component manufacture and assembly. As with most developments, a large fraction of the cost is determined in early design stages. With respect to high performance systems, the design choice determines not only the nominal performance but also the sensitivity to manufacturing tolerances. Tight tolerances mean high costs and in order to deal with tolerance induced performance degradation assembly of high-quality optical systems relies heavily on compensation strategies such as alignment. The development of optical systems hence exhibits a large amount of manual

---

* According to the Rayleigh criterion describing diffraction limited performance and distance referring to the optical path length [RAYL79].

assembly, making assembly a costly and often time consuming production step. Automation does not play a significant role because the produced lens designs change frequently and because assembly procedures requiring iterative adjustments are slow and difficult to automate [BECK05]. As a consequence, the optical industry is facing one of the dilemmas of production: highly specialized and complex small scale production with the need for scalable production technologies that enable the exploitation of economies of scale [SCHU07].

It is the goal of this work to contribute to solving this problem by investigating the application of a non-iterative assembly method which can compensate tolerance effects during the production of high-quality optical systems.

The work targets lens designers and optical engineers, providing the necessary tools to analyze and evaluate the applicability in early design stages as well as methods to successfully develop suitable optical systems. Most importantly, the performance of optical systems as a function of optical design, manufacturing tolerances and assembly method must be accurately predicted such that a cost-benefit analysis can be conducted.

1.1 State of the art – Developing high performance optical systems

Imaging optical systems have been designed and developed for a very long time, first through experimentation and crafting later by deliberate mathematical design. Calculation is based on Snell's law of refraction and the law of reflection which is sufficient for most applications as long as the finiteness of the wavelength can be neglected [BORN99]. Approximations of geometrical optics, explaining aberrations, the deviation of rays from perfect point imagery, made first performance increases possible.

The most important aberration theory is credited to Seidel [GROS07]. But more advanced theories for higher order errors [BUCH68] and asymmetric deviations exist [THOM80]. Since the 1980s the use of personal computers has further increased design performance. Following the works of Baker and Feder [FEDE62] computer-based optimization of a weighted error function (merit function) combining optical performance and boundary conditions is employed to maximize the performance. The merit function MF describes the weighted (w_i) difference of a design expressed by design functions f_i of parameters $\underline{x}$ to an ideal system state [GROS07].

1.1
$$MF(\underline{x}) = \sum_i w_i \cdot [\tau_i - f_i(\underline{x})]^2$$

The merit function contains targets τ_i for aberrations as well as first-order properties and boundary conditions such as geometrical restrictions. Sometimes the merit function is normalized by the sum of the weights.

Minimization of the merit function optimizes the performance and is based on nonlinear search algorithms including steepest descent, conjugate descent, Damped Least Squares (DLS) and Newtonian methods, finding local optima in the vicinity of a starting point [SMIT92]. Locating the global optimum of the merit function is an ongoing subject of discussion [KUPE93] and many optical design programs have implemented proprietary solutions. Algorithms are based on stochastic starting points, evolutionary and genetic search or simulated annealing [BASS09]. A systematic method to locate all different local optima by generating saddle points and subsequent local optimization with DLS is proposed by Bociort et. al. [BOCI05,VTUR09].

Despite the heavy use of computers to calculate and optimize optical systems, lens design is sometimes regarded an art [SHAN97]; only few design methodologies exist and they are rarely reported on. Intuition, experience and sometimes trial and error appear to dominate the development process to a large degree. As a consequence, optical systems are rarely developed from scratch. More typical, especially in longstanding companies, existing designs are modified and adapted to new specifications. Basis for these developments are lens databases, the patent literature or company own solutions. Haferkorn is among very few people describing a systematic process of optical system synthesis [HAFE84].

In addition, lens designers developed design strategies that make systematic use of aberration theory that may even be used to generate start designs for computer optimization. The design process can be divided into four stages: design choice, determination of powers and materials, shape adjustment and reduction of residual aberrations [SMIT00]. Kidger uses aspherical surfaces during preliminary design and suggests starting with monochromatic aberrations before correcting chromatic aberrations [KIDG01,KIDG04] while Shafer highlights the importance of aberration theory and aplanatic and concentric surfaces [SHAF80]. The methods are experience-based and general guidelines. Designing optical systems requires control over dozens of variables and even with today's optimization programs no design procedure surely leading to optical systems of the desired performance exists.

While optimal design performance is very important, the actual performance of optical systems is largely determined by manufacturing and assembly tolerances. Thanks to sophisticated tolerance analyses, the performance and yield of assembled systems can be accurately predicted. The importance of such analyses and careful tolerance assignment has grown immensely. Smith describes how deviations based on wavefront changes can be summed up to form a worst case or statistical estimate [SMIT85]. Adams provides statistical methods to predict the tolerance effect based on Gaussian tolerance distributions, including compensation [ADAM87]. Older publications develop analytical tolerance analysis models to reduce computational effort [KOCH78,PINT80]. Today, sensitivity analysis, Monte Carlo analysis and differential ray tracing analysis are the most important types [PERI05].

Traditionally, optical system design followed a strict sequential procedure: system layout, lens design and optical engineering [KING83]. While the system designer laid out the entire project and took account of specifications and cost, the lens designer refined the optical design before the optical engineer would assign tolerances, prepare drawings of mechanical parts and initiate manufacture and purchase.

Realizing the importance of tolerances on performance and cost, design for manufacturing strategies have emerged [KIDG04]. The strategies are mostly best-practice rules. Optimizing the optomechanical design to reduce tolerance effects [MARG99] and cost optimal tolerancing [YOUN01,WILL92] are some of the more systematic approaches. Including tolerance sensitivities in the optical design optimization is a major step towards an integrated product development. It aims at reducing the cost and increasing the robustness of a lens during design rather than finding a mechanical design that will enable the system to work [GREY70,YOUN06].

While the optical system design procedure has become more intertwined, the number of technical disciplines involved in the design process has also increased. This is because today's optical systems are characterized by a close interaction of optical, mechanical and often electronic components as well as an increased number of manufacturing alternatives [BLIE08]. As a result, the development process is often highly iterative and changes in the different phases result in a process of adaption that is dominated by the correct analysis of tolerances, environmental influences and their impact on system performance. Recently, simulation tools for the analysis of optical systems are brought closer together. Computer-aided design of mechanical parts is being connected to optical simulations and finite element analyses of structural and thermal behavior is combined with ray-tracing calculations to further increase performance [DOYL02].

Assembly, comprising the steps of joining, handling, alignment, test and additional processes is frequently one of the most important and costly production steps and responsible for a large fraction of added value [LOTT06]. This is particularly true for precision assembly with complex assembly routines. Of the different assembly steps, compensation is of particular interest for precision manufacture and very common in optics. Compensation is the ability to reduce errors induced by a set of parameter perturbations by (another) set of parameters.

Classical compensators in optical systems are air spaces and centration of lenses or lens groups to adjust symmetrical and asymmetrical aberrations such as on-axis coma, spherical aberration or distortion [GROS07]. Less frequently, rotation of lenses (clocking) is used to reduce on-axis astigmatism [GROS07]. Other errors can be compensated if parameters are measured prior to the assembly and air spaces or curvatures re-optimized [SCHL96]. Common examples include adapting a system to measured radii or glass melt data.

While adjustments moderately increase cost and require additional mechanical elements, reworking designated surfaces does not, but is only considered for very high-performance lenses [SCHL96]. Other related methods include centering of lenses and assemblies in mechanical cells using lathe centering or automated bonding [BLIE08]. Interferometric characterization and subsequent adaptation of the computer model to yield measured results can be used to iteratively adjust parameters [STEP89].

With increasing complexity, systematic selection of compensation parameters becomes more important. The selection can be based on sensitivity analyses of Zernike polynomials decomposing the wavefront into orthogonal polynomials [GROS07] that are suitable to measure compensation. Chapman and Sweeny find the most effective set of compensation parameters through singular value decomposition (SVD) of the design parameters' second derivatives [CHAP98].

Selective assembly is an assembly strategy used to improve quality and reduce costs and is based on measuring and sorting components (or subassemblies) into tolerance groups and selecting parts of matching groups for assembly [WARN96]. Selective assembly can be used to employ otherwise non-adjustable optical parameters such as lens curvatures as compensators.

Kidger describes the selective assembly of toroidal surfaces during the combination of achromatic doublets [KIDG04]. Application is also described by Ray [RAY02] who suggests the combination of suitably deviating components for the assembly of photographic lenses as well as by Thorburn [THOR83] and Haferkorn [HAFE84]. Adams states that tolerance analysis of selective assembly is still an open question in optical tolerance analysis [ADAM87] and Kingslake points out that a large number of lenses is required for what he calls matching [KING78].

Recently, Latyev et al. have published a study on the selective assembly of microscope objectives [LATY09] and [LATY10]. Research focuses on the adaptive and selective assembly method, developed by Zocher [ZOCH85] featuring a feedback loop into component manufacture and optimization of tolerance classes [GÖRS99] to avoid mismatch.

To summarize, non-iterative compensation methods which are most suitable for automation are rarely applied to optical systems. Infrequently, selective assembly is employed to increase performance but the usability for small quantities is limited and the method is not widely accepted. Multiple reasons can be identified:

- The method is not systematically conducted, but applied in a trial and error manner as described e.g. by [RAY02]. As a consequence, the performance increase remains uncertain.

- Interrelations between system parameters in optical systems are sometimes complex and require delicate and precise compensation of multiple parameters. Selective assembly is perceived as a strategy limited to simpler problems not suitable to solve optic-specific needs.

- A tolerance analysis concept for selective assembly, predicting the as-built performance does not exist. Hence, optical engineers do not know how to tolerance a design for selective assembly and cannot estimate a possible cost benefit.

- Lens design methods considering compensation methods are lacking. Most designs are therefore unsuitable for selective assembly and the full potential of selective assembly remains unused.

1.2   Goal and outline

In order to resolve the issues pointed out in chapter 1.1, the general idea of selective assembly will be taken up in this work but modified to better suit the needs of optic development. Instead of classifying components into tolerance groups, components are individually combined. Such individual component selection is unknown in the optics literature but is regarded a formidable solution for smaller production series and interrelated performance functions. In order to distinguish the method from selective assembly, the term combinatorial assembly is used hereafter.

The goal of this dissertation is to develop combinatorial assembly as a systematic method and employ it in the best possible way in order to facilitate cost-efficient production of high performance optical systems. Finding a cost-efficient approach during the conception and design of an optical system generally requires an iterative approach (Figure 1) that breaks up the traditional sequential development sequence.

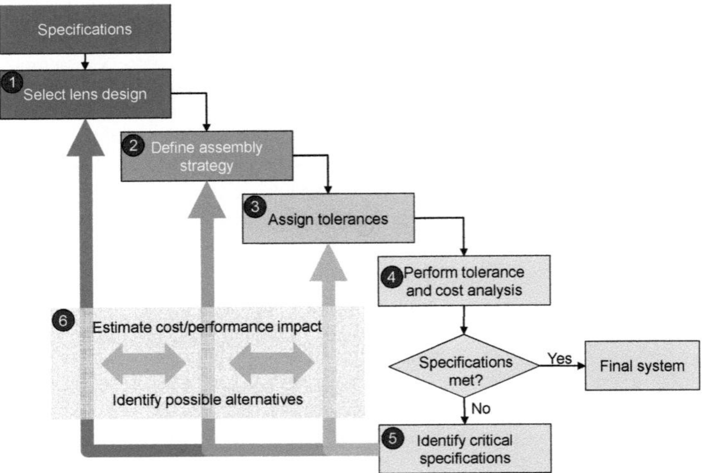

Based on given performance specifications a lens developer will derive an initial design and proceed to selecting an assembly strategy before he defines tolerances and analyses the performance (step 1-4). In a first trial, assembly without compensation and a simple set of tolerances will be tested. If cost or performance requirements are not satisfied, critical specifications are identified and possible changes discussed (step 5). Evaluating different options in the next step (step 6) is a classical engineering process based on intuition, experience and a good amount of systematic trial and error. After changes in lens design, assembly strategy and tolerances have been decided on the process starts again until a satisfactory result is obtained.

While the iterative approach is very generally applicable, this work expands the possibilities of the lens developer. New is the possibility to employ combinatorial assembly as an alternative assembly strategy (step 2) and analyze a systems performance with a specific tolerance analysis (step 4). New is also that combinatorial assembly requires an integrative discussion of its application as well as lens design and tolerance changes (step 6).

One of the goals is therefore to provide lens designers and optical engineers with suitable tools and knowledge to elaborate the application of combinatorial assembly. This requires the development of a systematic method to apply combinatorial assembly as well as simplified models, tolerance analyses and design methods.

Following a brief introduction to manufacturing tolerances, their effects on optical system performance as well as methods to analyze and assign tolerances this dissertation will

- develop the principle of combinatorial assembly including a formal description of the selection process. Different optimization methods delivering optimal component combinations are implemented, compared and their potential applications stated (chapter 3.1 and 3.2)

- formulate a systematic approach to selecting optical components for combinatorial assembly (chapter 3.3).

- expand tolerance analysis methods to predict the performance of combinatorially assembled systems and strategies to assign tolerances (chapter 4)

- contribute design methods to find optical designs optimized for the application of combinatorial assembly increasing possible error compensation and reduce tolerance requirements (chapter 5).

Finally, applications of combinatorial assembly are presented in chapter 6. The examples are carefully chosen to demonstrate the versatility of combinatorial assembly and its benefits in reducing errors of different orders.

2 Fundamentals of tolerances in optical systems

2.1 Manufacturing tolerances of optical components

A variety of processes for the fabrication of lenses and mirrors exist today [BLIE08] and despite the complexity of manufacturing optical and mechanical components, very high precision can be achieved. Nevertheless, every manufacturing process along the production chain exhibits natural fluctuations resulting in (mostly) minor imperfections impacting the performance.

Manufacturing deviations may be classified into regular and irregular errors. Whereas regular errors can be directly expressed as deviations of constructional parameters, irregular errors are described on a statistical basis. Figure 2 gives an overview of common manufacturing errors as defined in ISO 10110 [DIN00] and the US MIL-PRF-13830B [MIL97].

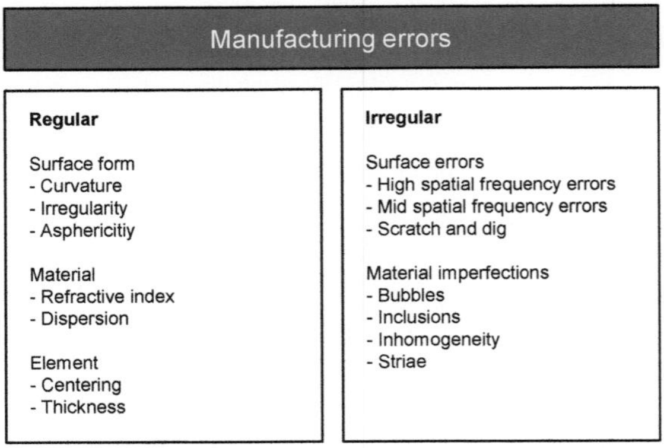

Figure 2
Manufacturing
errors of optical
elements classified
into regular and
irregular errors

Manufacturing errors	
Regular	**Irregular**
Surface form - Curvature - Irregularity - Asphericitiy	Surface errors - High spatial frequency errors - Mid spatial frequency errors - Scratch and dig
Material - Refractive index - Dispersion	Material imperfections - Bubbles - Inclusions - Inhomogeneity - Striae
Element - Centering - Thickness	

For the discussion of combinatorial assembly only regular errors are of interest while the treatment of statistical errors, coatings and stray-light is left to the literature [GROS07].

2.1.1 Description of regular manufacturing errors

The regular deviations of optical components as shown in Figure 2 are differentiated into material properties, element tolerances and surface form variations. In addition, opto-mechanical tolerances resulting from mechanical assembly will be discussed.

Material properties: The refractive index of glasses and its variation with wavelength λ, the dispersion, are subject to the chemical composition and manufacturing processes. Dispersion is best described by the formula developed by Sellmeier, a polynomial formula based on physical properties of glasses [SING06].

2.1

$$n_{Sellmeier} = \sqrt{1 + \frac{B_1 \lambda^2}{\lambda^2 - C_1^{\,2}} + \frac{B_2 \lambda^2}{\lambda^2 - C_2^{\,2}} + \frac{B_3 \lambda^2}{\lambda^2 - C_3^{\,2}}}$$

The coefficients B and C are the Sellmeier coefficients and determined by glass manufacturers from refractive index measurements at multiple wavelengths. While very accurate and almost always used during simulation, another dispersion formula is frequently employed in aberration theory and specified by glass manufacturers and during tolerancing: Abbe's v-number.

2.2

$$v = \frac{n_d - 1}{n_F - n_C}$$

The Abbe number relates the refractive index at a central wavelength (Fraunhofer d-line at 587.6 nm) to the dispersion over a range of wavelengths, typically from 486 nm to 656 nm (line F and C). Hence, the Abbe number depends on the refractive index at the central wavelength as well as on the change of the refractive index over a range of wavelengths. As both n_d and v are subject to tolerances, the index change far away from the d-line is much larger than for the central wavelength.

Element tolerances: Optical elements (e.g. simple or cemented lenses) are defined by the geometrical relations of optical and mechanical surfaces (Figure 3, left). Hence, tolerances for tilted and de-centered surfaces are either given with respect to the mechanical axis of an element or between optical surfaces.

Figure 3
Element tolerances
(left) and surface
tolerances (right)

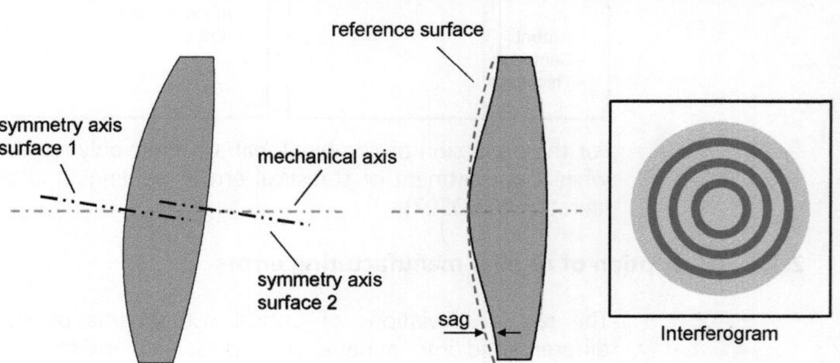

For arbitrary surface shapes tilt and decenter of the surface axis are given with respect to a mechanical reference. In the special case of two spherical surfaces a unique axis of symmetry for both surfaces, the optical axis, can always be found and related to the mechanical axis. The alignment of this optical axis to the mechanical axis describes the centration of optical components [GROS07], probably the most important element tolerance, and in general a vector quantity. In addition, element thickness is defined as the distance of surface vertices.

Surface form: While element tolerances relate optical and mechanical axes, the surface form error describes the difference between a real surface and a reference surface in alignment (Figure 3, center). The form error is measured along the optical axis and denoted surface sag. Interferometric measurements are commonly applied to determine the surface form deviations and led to the specification of interference fringes for tolerancing (Figure 3, right); one ring measuring the sag in units of half the wavelength of the test-light (typically 589 nm). First order form deviations are curvature differences as well as surface irregularity (toric surface) while higher order errors can be described using Zernike polynomials. Being a set of orthogonal polynomials, Zernike polynomials are particularly suitable for tolerancing purposes because different kinds of errors can be treated separately [GROS07].

Form errors of aspheres are often difficult to describe due to large differences to a reference sphere. The classical definition of the aspheric surface as given e.g. in [BRAU08] is not very suitable for tolerancing, as the coefficients do not have a representative meaning. Assigning tolerances is hence very difficult. Based on a similar idea than Zernike polynomials, Forbes developed a representation of aspheric surfaces that is particularly suitable for tolerancing [FORB08,YOUN09].

Opto-mechanical tolerances: During assembly, optical elements and mechanical mounts are brought together. As assembly typically requires a certain amount of play between lenses and mounts, lens positions are not necessarily very well defined. The resulting tolerances are decenter, tilt and axial shift of entire elements with respect to a mechanical reference axis. Two effects are of particular interest: a variation in element thickness can reduce an adjacent air space, the distance to the next optical surface. This is called an adjust air space, and which distance serves as an adjust needs to be carefully determined from the mechanical design. This is often difficult in preliminary design stages when neither the optical nor the mechanical design is fixed. In addition, a rotationally symmetric lens can roll on the mount such that a spherical surface will stay in place as it has an infinite number of symmetry axes [YODE08,DEWI05]. Hence, opto-mechanical tolerances require a great deal of attention and largely depend on the actual layout.

In order to limit the deviations of constructional parameters to an allowable range, tolerances are specified on engineering drawings. The specification of tolerances is regulated by the international ISO 10110. Table 1 summarizes typical tolerance values from commercial precision to a typical manufacturing limit. Higher precision can be achieved with specialized equipment.

Table 1
Tolerances of lens parameters for different levels of manufacturing precision
[OPTI10,BURG09]

	Unit	Commercial	Precision	Typ. limit
Optical				
Curvature/radius	%	±0.2	±0.1	±0.02
Curvature/radius	Fringes	5	3	1
Irregularity	Fringes	2	0.5	0.2
Center thickness	mm	±0.150	±0.05	±0.010
Scratch-dig	(MIL)	80-50	60-40	20-10
Surface sag	mm	±0.050	±0.025	±0.015
Index	%	±0.001	±0.0005	Melt Data
Dispersion	%	±0.8	±0.5	Melt Data
Mechanical				
Air space	μm	50	12	2.5
Centering	arcmin	6	1	0.25
Diameter	μm	100	25	6

The above tolerances hold true for grinding and polishing of lenses of 1-2″ diameter and have evolved over time. Surveys conducted by Plummer in 1979 [PLUM79] and Fischer in 1990 [FISC90] report tolerances that were approximately a factor of two larger than the values depicted in Table 1. The surface tolerances of polymer lenses fabricated by injection molding are typically larger by a factor of ten, while center thickness tolerances are comparable [WILL10]. For an in-depth discussion of optical and opto-mechanical tolerances the works of Yoder [YODE05], Fischer [FISC08], Adams [ADAM87] and Ahmad [AHMA97] are recommended.

2.1.2 The probabilistic nature of tolerances

Owing to statistical and systematic production errors, the components, manufactured to comply with specified tolerance limits, exhibit natural parameter deviations which can be described as a probability function φ. In general, these tolerance distributions depend on the manufacturing processes as well as on the desired tolerance limit. Typically, measured values cluster around the nominal value prescribed by the designer. Good approximations of the distributions are parabolic or Gaussian (normal) probability distributions. But sometimes other representations such as uniform and point-to-point are useful [GROS07]. Two examples are shown in Figure 4.

Figure 4
Tolerance
distributions:
truncated Gaussian
(left), Point-to-po nt
distribution (right)

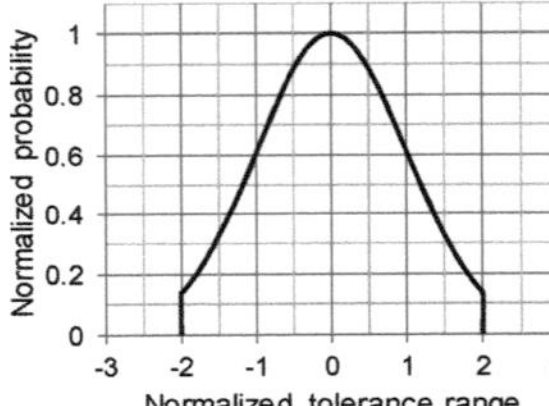

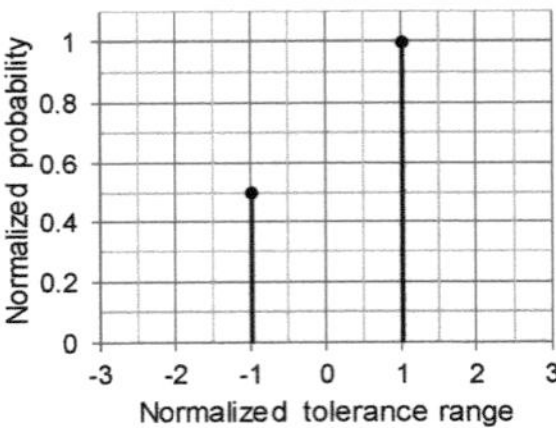

For tolerancing purposes, tolerance distributions are mostly approximated by Gaussian, truncated or shifted Gaussian distributions. Any dependence between tolerance distribution and prescribed tolerance range is usually neglected. That is, the shape remains constant, but the scale changes. The assumption of Gaussian distributions is particularly justified, if multiple tolerance distributions add up to a combined distribution (e.g. subassemblies or combined measures such as focal length). In these cases the sum of arbitrary probability distributions gradually converges to a Gaussian distribution on the basis of the central limit theorem [KARD05]. In practice, only a few Gaussian-like distributions need to add up to justify the assumption of a true Gaussian [SMIT85].

However, this is only true if tolerance distributions of distinct parameters are independent of each other. For most tolerances this is true, but there are notable exceptions. Surface quality and cleanliness are one such example [KREI10], while the dependence of center thickness and surface quality is another one. High surface quality demands longer processing times which in turn reduce center thickness. Hence, the according tolerance distribution is shifted to the lower end of thickness tolerances. While very useful and convenient for numerical analyses, great care is required in assuming an analytical tolerance distribution.

2.1.3 Cost of tolerances

If the cost of fabrication would not depend on tolerance limits, tolerances would always be assigned as tight as possible. Obviously, this is not the case and if tolerances are tightened, costs show an exponential increase. Figure 5 shows data reproduced from a survey conducted by Fischer [FISC90] indicating the exponential increase and also showing large differences for the participating shops. Naturally, the costs depend on available equipment and personnel and thus need to be determined individually to accurately reflect a shop's capabilities.

Figure 5
Cost functions for
center thickness and
diameter tolerance
reproduced from
[FISC90]

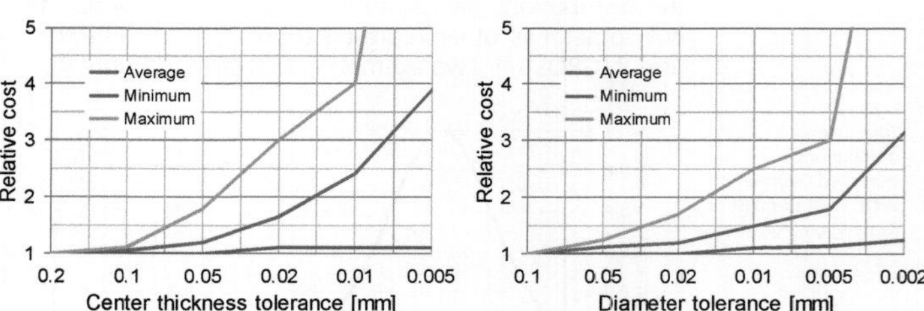

For better comparison, the cost functions are normalized to the base costs resulting from manufacture with standard tolerances. Surveys on manufacturing costs of optical components have been conducted by Plummer in 1979 [PLUM79], Fischer in 1990 [FISC90] and more recently by DeGroote Nelson et.al. in 2009 [DGRO09] while a cost model applicable to polymer lenses has been developed by Mäkinen and Nollau [MÄKI10].

Early cost models have been developed by Willey [WILL84,WILL92] assuming a simple inverse relation of cost and tolerances. Recent cost models are based on an exponential fit to approximate the data more accurately [DGRO09]. Other cost models feature step functions that may arise if more than one manufacturing technology is available [GROS07]. The relative cost functions of regular errors as published by DeGroote Nelson are given in Table 2.

Table 2
Relative cost
functions of selected
tolerances [DGRO09]

Tolerance	Unit	Relative cost factor
Power	µm	$1.2 \times (\text{Power})^{-0.4}$
Irregularity	µm	$(\text{Irregularity})^{-0.2}$
Center thickness	mm	$0.6 \times (\text{Center thickness})^{-0.3}$
Mean index	-	$0.06 \times (\text{Mean index})^{-0.4}$
Abbe number	-	$0.04 \times (\text{Abbe number})^{-0.7}$
Air space	mm	$0.6 \times (\text{Air space})^{-0.3}$
Element tilt	deg	$(\text{Element tilt})^{-0.2}$
Elem. decenter	mm	$0.6 \times (\text{Elem. decenter})^{-0.3}$

On average, going from standard tolerances to precision tolerances increases the cost of manufacture by 30%. The cost increase to achieve the manufacturing limit is substantially larger (compare Figure 5), making it well worth to think about cost-optimal tolerances.

2.2   Effects of errors on system performance

Calculating the effects of manufacturing and assembly deviations on the performance of optical systems is the primary interest of an optical engineer in order to determine a system's quality. While Maxwell's theory provides a comprehensive explanation for the propagation of electromagnetic waves, simpler theories are used to determine performance. Mostly geometrical optics is used to calculate how light passes through an optical system. This gross simplification is justified as optical wavelengths are much smaller than lens dimensions [HAFE03], and as limited as geometrical optics might seem, it turns out that it is not: diffraction at a large aperture can be described and wavefronts calculated, as can be polarization effects or scattering phenomena [TRÄG07].

For more than a century, optical systems from photographic lenses to lithography objectives have been designed and analyzed using geometrical optics. Unlike other engineering simulations exhibiting comparatively large uncertainties, real ray-tracing is very accurate and has been proven extremely successful. Typical properties that can be analyzed using geometrical optics are basic system parameters (first-order) and optical performance (higher order) as tabulated in Figure 6.

Figure 6
Optical specifications
as provided by
[FISC08]

Basic system parameters (First-order properties)	Optical performance (Higher order properties)
- Object/image distance - Total track - Focal length - F-number or numerical aperture	- Encircled energy - Modulation Transfer Function - Strehl ratio - Wavefront error - Distortion - Field curvature

While basic system parameters are often less critical, as the allowable variation is comparably large, optical performance needs more attention and will be the focus of the following discussion.

2.2.1   Optical performance measures

Imaging performance of optical systems is judged against perfect point imagery: rays that originate at a point in object space meet at a single point in the image space. This corresponds to spherical wavefronts which are concentric about the image point (see Figure 7). Real systems deviate from perfectly spherical wavefronts and tolerances disturb the wavefronts even further.

In order to evaluate an optical system, discrete field points, wavelengths and possibly multiple configurations are exemplarily evaluated. Measures that are used to determine the performance are either based on rays or on wavefronts that are determined by tracing rays through an optical system. Optical path difference (OPD) and ray fans are probably the most important measures during optical design, while customer specifications are more often expressed in terms of resolution, diffraction limitation or spot size. Ray fan plots are traditionally employed to judge imaging performance. The plots show the lateral deviation of ray intersections to a reference in the focal plane as a function of the pupil coordinate. Integrating the ray fan delivers the Optical Path Difference (OPD) representing the difference between a real wavefront and a reference sphere that would generate a perfect point image (Figure 7).

Figure 7
Ray paths through
an optical system
(meridional view)
with ray angles u
and ray heights y

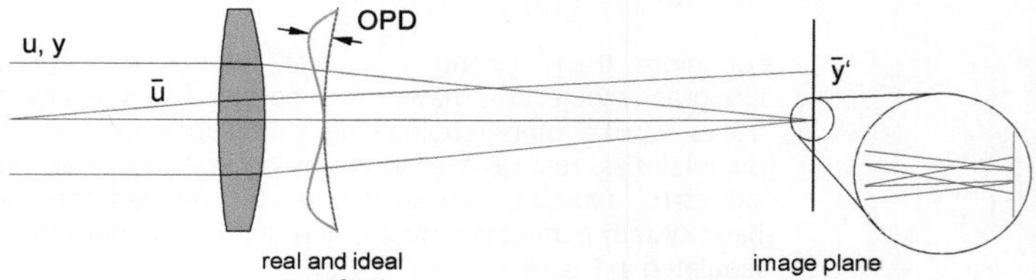

Usually, the reference sphere is located in the exit pupil [ZEMA10]. From the wavefront the Point Spread Function (PSF) can be calculated as a phase-right sum (Debye Integral) of rays emerging from the wavefront and accounting for the diffraction that occurs at the exit pupil. The ratio of theoretical peak intensity to actual peak intensity of the PSF, the Strehl ratio, is another frequently employed measure [MAHA91]. Resolution and contrast, especially interesting for imaging applications, can be evaluated using the Modulation Transfer Function (MTF), the real part of the Fourier transform of the PSF [SALE91]. For certain laser applications the encircled energy can be interesting. For a field point with spherical aberration balanced by defocus Figure 8 shows ray fan and OPD as a function of normalized pupil coordinates PY for the meridional plane as well as the PSF.

Figure 8
Common image
performance
measures from left
to right: ray fan,
OPD,PSF

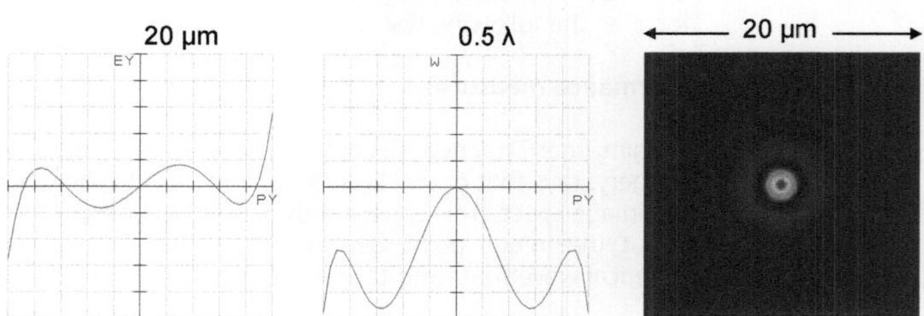

Diffraction limited systems play an important role in today's imaging optics and laser applications. A system is said to be diffraction limited, if the peak to valley OPD is lower than a quarter of the wavelength (the Rayleigh criterion) [RAYL79], the root mean square (RMS) wavefront error below 0.07 times the wavelength or the Strehl ratio larger than 0.8 [SMIT92]. These values are taken as guidelines for general purpose applications. However, some applications such as microlithography or astronomical telescopes require even higher performance.

Errors of optical components and assembly manifest themselves in deviations from the nominal performance. In order to gain deeper understanding a classification of independent types of errors provides useful. Two concepts which are in common use are a) Zernike polynomials, describing the wavefront as an orthogonal set of independent errors and b) aberration theories of different orders, explaining how aberrations develop. While the fundamental properties of optical systems are based on a first order approximation of Snell's law of refraction (Paraxial or Gaussian theory) substituting the sine by its argument, higher order approximations result in optical aberrations (from lat. *aberratio*, deviation). Hence, aberrations are departures of the optical performance from paraxial theory and their systematic analysis, description and calculation is called aberration theory.

2.2.2 Aberrations in centered optical systems

Seidel is credited to be the first who extended the Gaussian theory and mathematically described aberrations providing a sound understanding of geometrical optics. Expanding the sine in Snell's law of refraction into a Taylor series he used the first and third-order terms to describe optical aberrations. From this theory, five monochromatic aberrations can be identified and are known as spherical aberration, coma, astigmatism, distortion and field curvature. In addition two chromatic aberrations, longitudinal and axial chromatic aberrations exist [GROS07].

Seidel aberrations are calculated from paraxially traced marginal and chief rays and expressed as a sum of surface contributions. Every surface has a contribution to the aberrations and the individual surface contributions can be expressed as aberration coefficients [GROS07]. During lens design, the sum of the individual contributions of separate surfaces will be reduced by balancing positive and negative contributions against each other. Because third-order theory is only an approximation, a full correction includes compensation of aberrations of different orders. Designing a system that is perfectly corrected for third-order aberrations is rarely a good system. Even systems corrected for all third- and fifth-order aberrations can be far from optimal yet good control over third order aberrations is mandatory in order to deal with higher orders [SHAF94].

The following figure shows the refraction of upper and lower marginal rays at a single surface. Let n denote the refractive index, u the ray angle and h the marginal ray height as well as i the angle of incidence.

Figure 9
Ray data definitions
to compute
aberrations

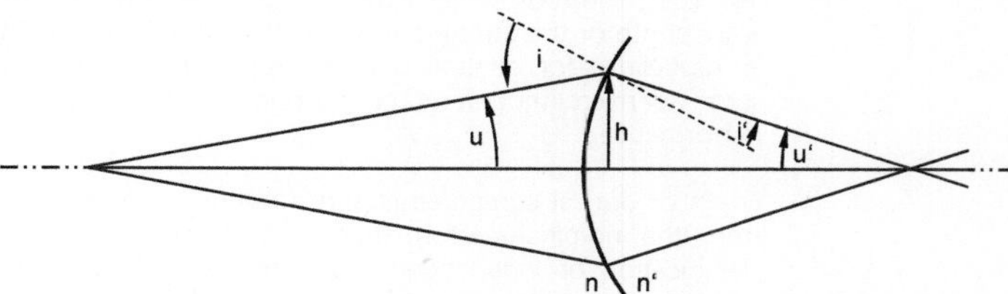

Let also identify quantities after refraction with a prime and the chief ray with a bar, while all other quantities refer to the marginal ray. With $H=nyu$ the Lagrange invariant, y the object height and $A=ni$ (again, barred quantities refer to the chief ray) the aberration coefficients s_i then read as:

2.3

Spherical

$$s_1 = -A^2 \cdot h \cdot \left(\frac{u'}{n'} - \frac{u}{n} \right)$$

Coma

$$s_2 = -A \cdot \overline{A} \cdot h \cdot \left(\frac{u'}{n'} - \frac{u}{n} \right)$$

Astigmatism

$$s_3 = -\overline{A}^2 \cdot h \cdot \left(\frac{u'}{n'} - \frac{u}{n} \right)$$

Field curvature

$$s_4 = -H^2 \cdot c \cdot \left(\frac{1}{n'} - \frac{1}{n} \right)$$

Distortion

$$s_5 = -\frac{\overline{A}}{A}\left(s_3 + s_4 \right)$$

The balance of aberrations achieved during design is disturbed by tolerances of curvatures, materials, thicknesses and air spaces as the paths of rays through the system changes. As can be seen from the formulas, the changes of the aberrations induced by parameter variations are of different orders. In addition, the change of a parameter can influence the aberration contribution of multiple surfaces as ray angles and heights at following surfaces change.

A few interesting insights can be drawn from the aberration equations. If the angle of incidence of the marginal ray i becomes zero, spherical aberration and coma disappear; the surface is concentric. Similarly, if the angle of incidence of the chief ray vanishes, coma and astigmatism are zero. Finally, if $\Delta(u/n)$ is zero, only field curvature remains. This is called the aplanatic case and many lens designs make use of this type of surface. The aberrated wavefront W can be determined with the help of the aberration sums S of the surface contributions s_i in the following way [GROS07].

2.4

$$W = \frac{1}{8}S_I \rho^4 + \frac{1}{2}S_{II}\eta\rho^3 \cos\varphi + \frac{1}{2}S_{III}\eta^2\rho^2 \cos^2\varphi + \frac{1}{4}\left(S_{III} + S_{IV} \right)\eta^2\rho^2 + \frac{1}{2}S_V\eta^3\rho^1 \cos\varphi$$

ρ, η and φ are normalized pupil coordinates, field coordinates and azimuth angle respectively. The weighted coefficients are often denoted W_{ijk}, where i,j,k denote the exponentials of ρ, η and $\cos\varphi$. Theories for fifth [BUCH68] and even higher order aberrations [ANDE80] have been developed and contain aberrations such as oblique spherical or elliptical coma [KIDG04]. In contrast to Seidel aberrations higher order aberrations are often induced by aberrations of another surface and difficult to interpret [SHAF89].

2.2.3 Aberrations in non-centered optical systems

Many of the most relevant performance degradations are induced by errors disturbing the axial symmetry of a system. The effects of decentered optical elements have been studied by Rimmer, Hopkins and Tiziani, and many others [RIMM70,HOPK66].

Vector aberration theory is particularly suitable for the description of asymmetrical aberrations and has been developed by Shack, Buchroeder (third-order) and Thompson (fifth-order) [THOM80]. Central idea of the theory is that aberration types are similar to symmetrical aberrations but that aberration fields are decentered with respect to a reference axis. The new aberration fields are centered on a line connecting the center of curvature of individual surfaces and the pupil. By superposition of the individual aberration fields the total aberration of the system results (Figure 10, left).

Figure 10
Vector aberration:
superposition of
aberration centers
(left), compare
[ROGE06] and
determination of
incident angles
(right)

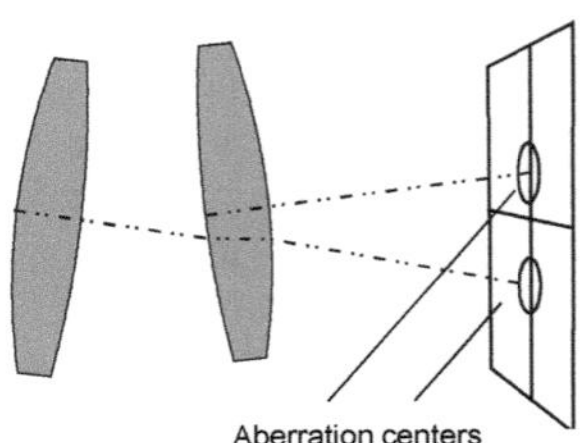

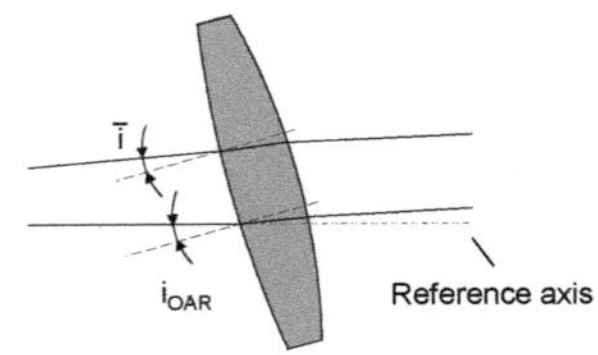

Central is the determination of the optical axis ray and in particular the angles of incidence i_{OAR} on individual surfaces. The optical axis ray is a ray along the original optical axis which is bent due to refraction at decentered and tilted surfaces. Calculation is based on determination of a generalized tilt parameter β describing the position of the surface. Equations are in the appendix (1.1).

For the third-order contribution the surface contributions to the wavefront (in polar form) as given in 2.4 is extended to a vector formulation and the field vector η replaced by η-σ [TURN92].

2.5

$$W_i = w_{040,i}\left(\underline{\rho}\cdot\underline{\rho}\right)^2 + w_{131,i}\left[\left(\underline{\eta}-\underline{\sigma}_i\right)\cdot\underline{\rho}\right]\cdot\left(\underline{\rho}\cdot\underline{\rho}\right) + w_{222,i}\left[\left(\underline{\eta}-\underline{\sigma}_i\right)\cdot\underline{\rho}\right]^2 +$$
$$w_{220,i}\left[\left(\underline{\eta}-\underline{\sigma}_i\right)\cdot\left(\underline{\eta}-\underline{\sigma}_i\right)\right]\cdot\left(\underline{\rho}\cdot\underline{\rho}\right) + w_{311,i}\left[\left(\underline{\eta}-\underline{\sigma}_i\right)\cdot\left(\underline{\eta}-\underline{\sigma}_i\right)\right]\cdot\left[\left(\underline{\eta}-\underline{\sigma}_i\right)\cdot\underline{\rho}\right]$$

The total wavefront contribution then results from summing the individual W_i of every surface. For coma, frequently one of the most important aberrations resulting from decentration tolerances, the following expression results:

2.6

$$W = \left(\sum_i W_{131,i}\right)\left(\underline{\eta}\cdot\underline{\rho}\right)\cdot\left(\underline{\rho}\cdot\underline{\rho}\right) - \left(\sum_i W_{131,i}\,\underline{\sigma}_i\cdot\underline{\rho}\right)\cdot\left(\underline{\rho}\cdot\underline{\rho}\right)$$

The first term is the total coma contribution of all surfaces for the symmetrical case, while the second term constitutes the asymmetric coma contribution. It depends on the vector $\underline{\sigma}$ which can be determined as follows:

2.7

$$\sigma_i = -\frac{i_{OAR,i}}{i_i}$$

Hence, it is the ratio of the angle of incidence of the optical axis ray and the chief ray angle of incidence for the unperturbed case. Again, sensitivity to tolerances can be expressed in terms of ray angles. Through superposition of individual and shifted aberrations, points in the image with vanishing aberration contributions, so called nodes, exist. The number of nodes depends on the field dependence of the ordinary aberration as given in 2.4. Hence, as symmetrical coma depends linearly on the field, asymmetrical coma is invariant with field size.

2.2.4 Zernike polynomials

Zernike polynomials were developed to describe wavefronts with circular apertures. As most optical systems are rotationally symmetric, their apertures are too, and so are the wavefronts. This suggests describing deformations of a wavefront by a set of orthogonal polynomials defined in polar coordinates. The Zernike polynomials Z are defined as follows [GROS05] and consist of a radial term R which depends on the normalized aperture radius r and of a term that depends on the azimuthal angle φ:

2.8

$$Z_n^m(r,\varphi) = R_n^m(r)\cdot\begin{cases}\sin(m\varphi) & for\ m>0\\ \cos(m\varphi) & for\ m<0\\ 1 & for\ m=0\end{cases}$$

The indices n and m represent the different orders and are such that $n\text{-}m$ is always an even number and $m \leq n$. Typically, only the first 36 polynomials are considered and different schemes to order them exist. The first polynomials in fringe indexing are piston error and tilt in x- and y-coordinate direction followed by defocus (Z_4) and the classical aberrations astigmatism (Z_5 and Z_6) and coma (Z_7 and Z_8).

Using the polynomials, a wavefront W can be decomposed into contributions of the different polynomials by calculating coefficients c_{nm} which determine the magnitude of a particular error.

2.9
$$W(r,\varphi) = \sum_n \sum_{m=-n}^{n} c_{nm} \cdot Z_n^m(r,\varphi)$$

The coefficients are thus a measure of the size of an aberration. Due to their orthogonality the Zernike coefficients of lower orders do not depend on higher orders. This is a great advantage in interpreting the aberrations of a decomposed wavefront. In the case of only third-order aberrations Zernike coefficients correspond directly to Seidel aberrations [GROS07].

2.3 Tolerancing of optical systems

The task of assigning a set of allowable deviations to system parameters in order to meet performance and cost demands is called tolerancing, or sometimes tolerance budgeting [SMIT85].

Starting with a tolerance analysis to determine the effects and their causes the process of assigning tolerances is often iterative and frequently alignment and test procedures need to be included in the tolerancing process [YOUN06]. At the interface between optical design, mechanical design, manufacture, assembly and test, tolerancing is not only required to meet specifications of cost and performance but also of manufacturability and of measurability. Obviously, a compromise is often necessary. The opposing quantities of cost and performance are most important for the following discussion.

2.3.1 Tolerance analysis

Tolerance analyses are used to evaluate the effect of particular tolerances on system performance, or, more importantly, to estimate a system's final performance. The final performance is sometimes referred to as the as-built performance and tolerance analyses need to account for manufacturing tolerances *and* assembly steps such as alignment. Performance criteria can be the above mentioned image quality criteria as well as constructional parameters and first-order properties. Most of the discussions in the literature, however, focuses on imaging properties as these are usually critical for imaging systems [YOUN06].

During **sensitivity analysis** parameters are successively altered to their upper and lower tolerance limits and the change in performance q is recorded. No limitations to the type of criteria need to be made. Always only one parameter is disturbed and a so called change table of the recorded performance change is created. If x_i denotes the nominal value of parameter i and t_i denotes the tolerance change of that parameter, the sensitivity S of a parameter i is given as [ADAM87].

$$S_i = \frac{q(x_1,...,x_i + t_i,...,x_n) - q(x_1,...,x_i,...,x_n)}{t_i} \qquad \text{2.10}$$

Thus, sensitivity is defined as a relative change and most suitable for the analysis of linear changes. For linear and strictly symmetrical perturbations (e.g. centration in symmetrical systems) the performance change due to positive and negative perturbations is naturally equal while in other cases different changes might be produced. In order to assign each tolerance a single sensitivity value, positive and negative changes are sometimes combined to yield an average [ZEMA10]. However, measures with nonlinear behavior are not particularly suitable for sensitivity analysis.

While in theory, a change table for every specification can be set up, this would be very difficult to analyze and is rarely discussed in the literature. Instead, discussion focuses on imaging quality as other performance criteria are typically far less of a concern. For the application to imaging quality (e.g. aberrations) the performance at multiple field points, wavelengths or configurations of the optical system is combined into a single performance criterion. Very typical is a combination of wavefront errors in the form of a merit function. Others use only the most sensitive or critical field point [SMIT85].

Calculating sensitivities provides good insight into the effects of single perturbations. In order to predict system performance after manufacture and assembly it is necessary to consider the effects of all tolerances simultaneously. Two kinds of estimations are frequently used: the worst case estimate and the root sum square (RSS) estimate [SMIT85]. Worst case estimation assumes independence of tolerance effects and linear superposition. The calculated changes $S_i t_i$ are summed up to yield the worst case estimate q_{WC}.

$$q_{WC} = \sum_i S_i t_i \qquad \text{2.11}$$

Though very intuitive, this procedure has a major drawback. Tolerance effects are frequently overestimated as cancellations and effects relating to the central limit theorem are not considered.

The root sum square estimate q_{RSS} avoids statistical overrepresentation. Instead of linearly adding effects the square root of the sum of squared effects is taken as the estimate. The likelihood of every tolerance being at the limit is thus better represented [SMIT85].

2.12

$$q_{RSS} = \sqrt{\sum_i [S_i t_i]^2}$$

A statistically exact analysis (e.g. performance distribution) taking every tolerance into account can be obtained by the **Monte Carlo method**, treating tolerance analysis as a probability experiment. This is used for a variety of statistical analyses in optical research [FRIE80].

For a given optical system and known probability distributions of tolerances the estimated performance is calculated by generating a large number (e.g. >1000) of random systems. For every random system the parameter values are determined as random numbers according to the specified probability distributions of tolerances. Thus, every system parameter is disturbed simultaneously. Typically, it is assumed that parameters are independent of each other and commercial ray-tracing software treats Monte Carlo this way [ZEMA10]. Nevertheless, accounting for interrelations between parameter values would be possible. Concerning the number of Monte Carlo runs, only few guidelines exist. Gross recommends several thousand runs [GROS07] while simply comparing the results of different runs to determine a sufficient number of systems is probably the safest.

Because of the large number of systems which need to be calculated, the Monte Carlo method is a rather slow analysis. As an alternative, analytical methods predicting statistical probability distributions of performance measures have been developed, mostly to avoid computationally intensive methods such as Monte Carlo [PINT80,ADAM87,KOCH78]. The methods assume linear or quadratic relations of the performance criteria and develop the statistical mean, variance and higher moments for a given tolerance distribution – mostly Gaussian.

Differential ray tracing is based on the fundamental works of Rimmer [RIMM70] as well as of Hopkins and Tiziani [HOPK66], investigating wavefront changes induced by disturbed parameters.

Differential ray-tracing is based on linear approximations of the changes in ray position and ray angle. Differential ray-tracing is applied in paraxial ray racing and calculating Gaussian beams. As an approximation it is very useful for tolerancing as estimated performance changes are based on wavefront changes. Figure 11 illustrates this for a two-dimensional example showing the optical path length change for a tilted or decentered surface.

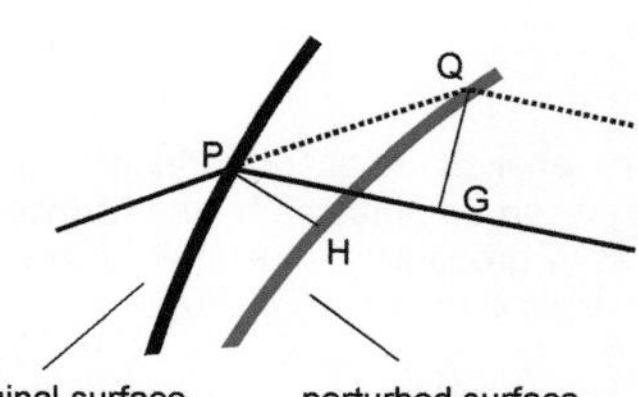

Figure 11
Optical path length
change due to
perturbed surface,
compare [RIMM70]

It emanates that exact differentials are not necessary for tolerancing purposes. Instead the change in optical path length

2.13
$$\delta_{OPL} = n\overline{PQ} - n'\,\overline{PG}$$

is determined by a change of the normal distance R between the ideal and perturbed surface.

2.14
$$R = \overline{PH}$$

If i and i' denote the original angles of incidence and exitance and n and n' the refractive indices before and after refraction the change amounts to:

2.15
$$\delta_{OPL} = \left(n\cos i - n'\cos i'\right)\cdot R$$

Using this formula, the differential ray-tracing procedure provides very useful for fast, yet accurate tolerance analysis. It is however limited to effects that directly relate to wavefront changes. For all of the above mentioned analyses, alignment procedures can be incorporated by evaluating a lens after compensation.

2.3.2   Tolerance budget and cost optimal tolerancing

The assignment of a tolerance budget aims at achieving the desired performance in the most economical fashion. Traditionally, meeting performance criteria dominated the assignment of tolerances and intuitive proportions were given to individual tolerances. Nowadays, cost based approaches have emerged [YOUN01]. While many other costs are fixed after design completion (e.g. number and material of components) tolerances can be adapted and their respective costs minimized.

Smith provides a scheme to assign tolerances based on sensitivity analyses. Using the RSS estimate the allowable tolerance degradation from a nominal system in terms of the optical path difference OPD_{tol} is determined as

2.16
$$OPD_{tol} = \sqrt{OPD_{spec}^2 - OPD_{nom}^2}$$

As the nominal OPD_{nom} is different for every field point, Smith suggests using the most critical as a reference to determine the tolerable change OPD_{tol} [SMIT85]. OPD_{spec} denotes the specification limit for the optical path difference, e.g. the Rayleigh limit. From a change table stating OPD changes due to tolerance deviations the RSS is used to estimate the performance change. Tolerances are then adapted to bring the estimated performance to the allowable limit.

Ideally, cost-optimal tolerance assignments are sought and Smith includes simple approaches in his method [SMIT85]. Adams systematically develops cost-optimal tolerances for linear effects taking performance limits into account by using Lagrangian multipliers [ADAM87]. Defining a new set of functions F that combine cost C and performance q the cost-optimal tolerances are found if the partial derivatives of F with respect to the tolerances t are all zero.

2.17
$$F(\underline{t}) = C(\underline{t}) + \lambda \cdot q(\underline{t})$$

Here, q is the performance boundary condition (e.g. target value of the estimated worst case) and λ the Lagrangian multiplier. For a worst case estimate and an inverse cost law, tolerances should be assigned in inverse proportion to the square root of the sensitivity to minimize cost.

2.18
$$t_i = \frac{1}{\sqrt{\lambda |S_i|}}$$

More exact results can be obtained using complex performance estimates such as the statistical RSS prediction as in equation 2.12. If resulting tolerances are outside their feasible limits, an iterative approach can be used setting their values exactly to the tolerance limit and finding new optimal tolerances for the remaining parameters [ADAM87]. The concept can be extended to multiple performance criteria (e.g. aberration types) but is limited to effects that relate to OPD [PERI05].

Youngworth and Stone propose a cost-optimal tolerancing method based on a quadratic approximation of a single function of merit MF around the design optimum $\underline{x}_0$:

2.19
$$MF = MF_0 + (\underline{x} - \underline{x}_0) \cdot \frac{\partial MF}{\partial \underline{x}} + \frac{1}{2}(\underline{x} - \underline{x}_0) \cdot \frac{\partial^2 MF}{\partial \underline{x}^2} \cdot (\underline{x} - \underline{x}_0)^T$$

Using this approximation to predict the performance, an automated, iterative approach can be used to determine cost-optimal tolerances [YOUN01]. A more rigorous approach to cost-optimal tolerancing involves Monte Carlo simulations instead of performance approximations as presented by Kehoe [KEHO10].

2.3.3 Compensating effects of errors

Compensation is the ability of a set of parameters (the compensators) to reduce the effects of errors produced by deviations of another set of parameters [GROS07]. Different terms are in current use: adjustment, alignment and compensation. While adjustment and compensation are often used synonymously, alignment refers to the process of centering the optical axes of lenses and mirrors. Typical compensators are air-spaces and lateral lens positions correcting spherical aberration and on-axis coma, respectively [GROS07]. Most frequently, back focus compensation is employed.

In order to perform compensation, the optical system is characterized after it is assembled and the compensators iteratively changed to minimize the errors. This is illustrated in Figure 12 (left) where parameter A is the compensator taking different values depending on the value of parameter B, subject to a tolerance distribution φ_B.

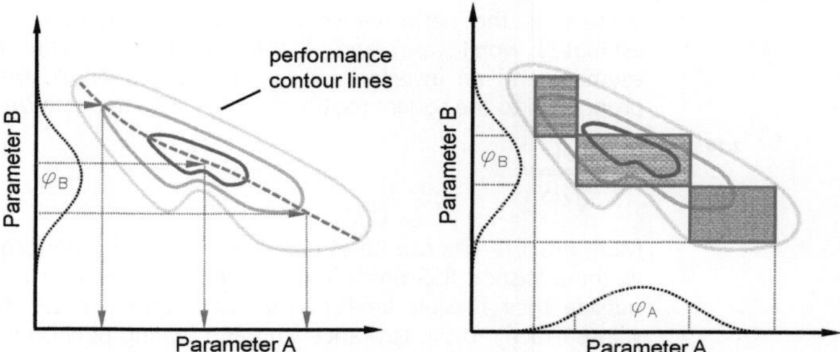

In order to compensate errors, suitable compensators need to be identified from the optical design. This is usually done with a sensitivity analysis determining the sensitivity to Zernike aberrations [GROS07]. Complicated assembly routines can be investigated with Zernike polynomials [WILL89] and the best set of compensators identified from singular value decomposition [CHAP98].

Selective Assembly is also a form of compensation but is non-iterative. Parameter values of different components are classified into tolerance classes depending on their parameter deviation as shown in Figure 12 (right). Components falling into corresponding classes are assembled [WARN96]. For selective assembly, the cumulated probabilities for every matching class must agree to avoid mismatch and strategies to dynamically adjust classes exist [GÖRS99].

2.3.4 Tolerance desensitization

Both cost and performance largely depend on the optical system a designer has developed. Tolerance desensitization aims at modifying optical designs as to make them insensitive to tolerance changes, particularly opto-mechanical deviations such as centration and tilt. Schematically, this is often represented as illustrated in Figure 13 showing a two-dimensional design space and the corresponding cross section.

Figure 13
Illustration of
lens design
desensitization

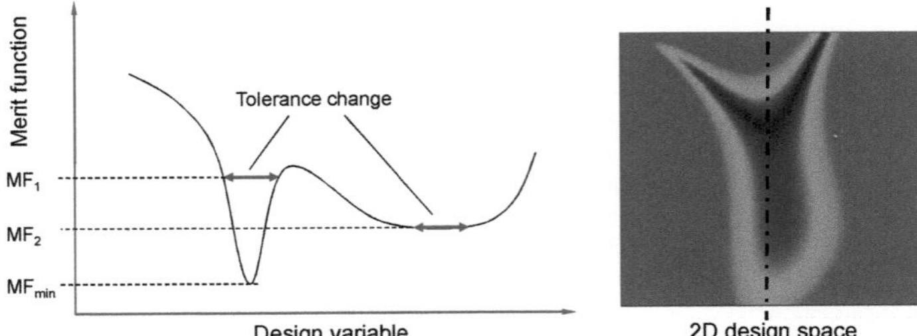

Of the different local optima of the merit function, a local design optimum with a large plateau of constant merit function values is less sensitive to parameter perturbations than a steep minimum. Often the as-built performance of less sensitive designs can be better, even if the nominal performance is worse ($MF_2 < MF_1$). The expression of stress, describing systems that are more susceptible to perturbations and characterized by large aberration contributions of single surfaces has first been formulated by Glatzel [KIDG04]. In these systems, higher order aberrations are typically large to counterbalance lower order aberrations. Keeping the performance level but reducing the sensitivity will yield robust and inexpensive systems. From manufacturing and cost perspective it is very sensible to include a system's sensitivity to tolerance deviations, and therefore costs, into the merit function as well. Additional targets τ_i are introduced into the merit function measuring the sensitivity of the design. Some use the sensitivity itself to desensitize systems [ZEMA10]. However, other measures have been found to be very effective in reducing the sensitivity during optimization.

Grey uses the differential ray tracing formula to estimate the sensitivity [GREY70]. Isshiki incorporates the change of ray angles due to refraction at optical surfaces into the design merit function, assuming that small angles of refraction correspond to small aberration contributions and small aberration contributions exhibit small changes due to tolerances [ISSH07,ISSH04]. Dilworth [DILW08] considers third-order aberrations, whereas predicted wave aberrations due to manufacturing errors have been included in the merit function by Dobson and Cox [DOBS95]. Optimizing multiple configurations of a lens design, each with small differences representing tolerance based deviations, has been adopted and modified by a number of researchers. Apart from the design system, additional configurations with disturbed parameter values are created and their parameter deviations linked to the nominal system. Optimization of a combined merit function including the performance of every configuration will then find a best compromise. Rogers uses only a small number of configurations but perturbs every parameter simultaneously [ROGE06]. Because the parameter deviations are randomly distributed a statistically representative set of configurations is established.

Global optimization techniques have been adopted to reduce tolerance sensitivities. Using a merit function based on the as-built performance, McGuire uses global optimization to find desensitized design forms [MCGU06]. Epple and Wang take an entirely different path and introduce new design means into the system to further reduce tolerance sensitivity. The use of aspheric surfaces proves to be a powerful means of reducing tolerance sensitivities. The advantage of this strategy lies in additional degrees of freedom during the design process allowing for better control of sensitivities.

Designing tolerance insensitive systems can also be achieved in a more design-methodological fashion. Shafer describes the modular generation of systems based on aplanatic and concentric surfaces and elements. As the third-order aberration contributions of spherical aberration, coma and astigmatism are zero for such elements the approach yields comparatively insensitive design forms [SHAF80].

2.4   Summary and conclusion

Manufacturing and assembly errors reduce the performance of optical systems and assigning tolerance limits plays a major role in optical system engineering, connecting lens design with manufacture and assembly. As such, performance and cost of an optical system are influenced by tolerances and the resulting effects need to be taken into account.

While calculation and estimation of effects on system performance due to tolerances can be performed with a high degree of precision, the assignment of tolerances poses more difficulties. Defining a cost-optimal tolerance budget is an ongoing topic in the research community as is the design of systems with optimum as-built performance.

Increasing the performance of optimized systems can be achieved by choosing a suitable lens design and by compensating undesired effects. The main method is alignment, which is difficult to automate and comparably slow. Therefore, the next chapter will develop combinatorial assembly as an alternative compensation strategy.

3 Combinatorial assembly of optical systems

3.1 Combinatorial assembly - a compensation strategy

As discussed in chapter 2 optical systems are characterized by a large degree of interrelated performance functions as well as nonlinear relationships between parameters and functions. This distinguishes optical systems from many other products, especially mechanical assemblies. While one of the fundamental principles of mechanical design is the separation of functions [PAHL06] this principle is inherently violated for optical systems where single surfaces influence multiple aberrations. The trend with optical (and mechanical) systems towards a large degree of system integration and the development of new and complex surface types lead to systems with strong interdependencies. The inherent complexity of optical systems can therefore lead to extremely tight tolerances and make compensation and alignment techniques during assembly necessary to guarantee performance.

3.1.1 Principles of combinatorial assembly

From a production management point of view, assembly is a transformation process [DYCK07]. Components are combined to subassemblies and subassemblies to systems, and the entire assembly can be divided into separate production processes such as joining of components, alignment or the selection and matching of components that will be described subsequently. A graphical representation of the assembly of a laser focusing lens is given in Figure 14.

Figure 14
Assembly graph of a
laser focusing lens

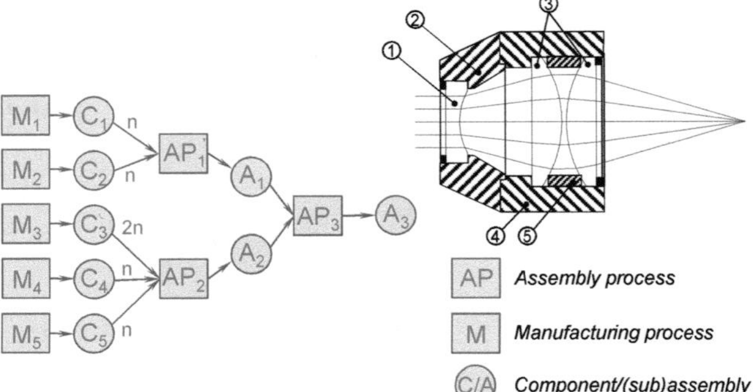

The lens consists of five different component types C_i. Components, subassemblies A_i and final systems are depicted as nodes, while rectangles represent assembly (e.g alignment) and manufacturing processes AP_i and M_i, respectively. A single system consisting of $m=5$ different types of components is transformed to a final system by successive assembly processes. If a series of systems is produced n_i components of every component type are used to assemble n systems as indicated in Figure 14. Note that in this example two lenses (C_3) are of the same kind and are thus represented by a single node with an according amount of twice the number of systems.

Every assembly process AP_i during the assembly of an optical system can be different. During classical assembly, parts and components are usually randomly chosen from a pool of existing parts and then put together. If necessary, alignment is employed to obtain higher performance. Most typical are alignment and adjustment procedures that involve lateral or axial movement of components. Sometimes selective assembly is applied to large volume production. Combinatorial assembly tries to close the gap and aims at using the maximum compensation potential by individually matching components. Figure 15 illustrates classical assembly, compensation, selective assembly and combinatorial assembly for a two-dimensional case with parameters belonging to two different components.

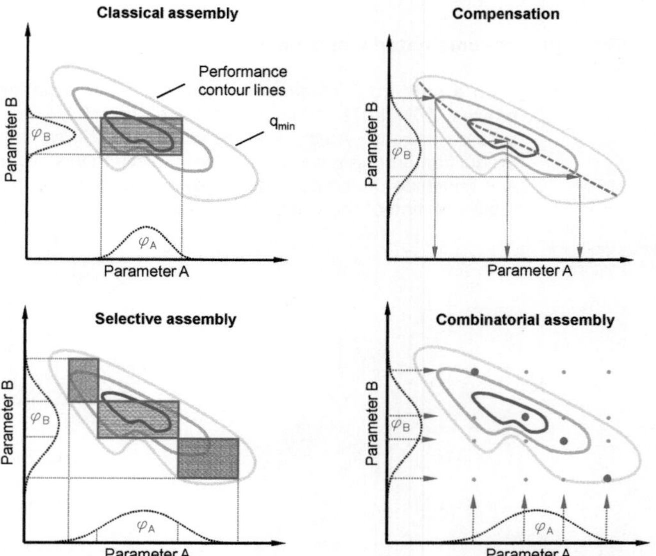

For classical assembly a tolerance region is spanned by component tolerance distributions φ. This region needs to be small to fit inside the desired limits q_{min}

of the performance function q, represented by its contour lines. Tolerances maximizing the yield are necessarily tighter than for the case of compensation where one parameter (here parameter A) can be freely chosen to optimize the performance. Selective assembly divides the tolerance distributions into classes spanning multiple sections to better approximate the performance function q.

As a new assembly method for optical systems, combinatorial assembly is similar to compensation and may be regarded a form of discrete compensation. The basic idea behind combinatorial assembly is to select optimal component combinations of a number of available components, here indicated by the arrows in Figure 15. As illustrated for four components of type A and four of type B, some combinations fall inside the performance requirement q_{min}, while other combinations do not. In selecting suitable combinations of components as illustrated by the highlighted dots in Figure 15, it is possible to increase the average performance and decrease the variance in performance of a production series (a set of systems) by avoiding worst-case combinations. A reduction of the variance is comparable to an increase of process capability of production processes, the ratio of the allowable range of specification to six times the standard deviation [SCHM10]. Tolerances can in turn be loosened, costs saved or high performance demands achieved.

Consider for example the two subassemblies A_1 and A_2 of the laser focusing lens in Figure 14. If during combinatorial assembly the selection of ten subassemblies is optimized in an attempt to reduce on-axis errors arising from misaligned modules the performance improvement as depicted in the histogram in Figure 16 can be obtained. The figure shows the probability $\Phi(q)$ for intervals of the on-axis error. The data is derived from simulations with a tolerance analysis tool that will later be developed and shows a reduction of the variance of on-axis coma for combinatorial assembly.

Figure 16
Histogram showing the performance improvement by combinatorial assembly in comparison to random assembly

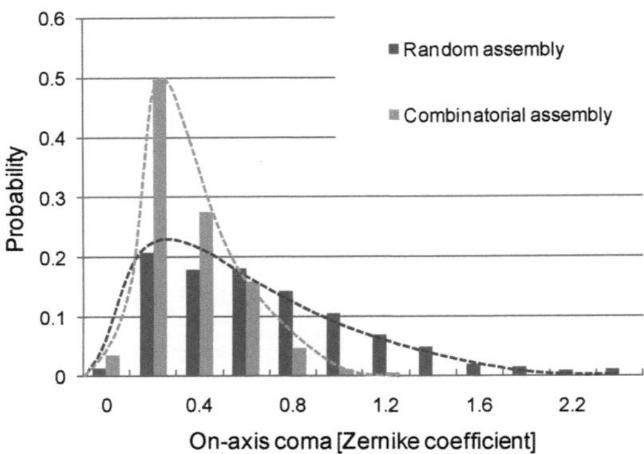

In theory, finding the optimal component configuration for combinatorial assembly can be achieved in a trial and error manner and, as a matter of fact, many practical problems are solved this way too. Different combinations of the available components are assembled and then tested until a satisfactory result is achieved [RAY02].

This is not very satisfying and to avoid blindly probing different combinations, a concept will be derived to determine the optimal selection of components prior to the assembly as shown in Figure 17.

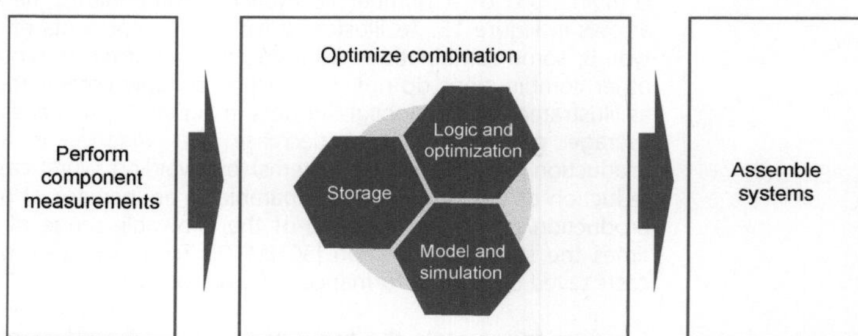

Components are first measured and the acquired information is stored in a database. Then, a physically exact model, accurately predicting the performance of assembled systems is employed to determine the optimal combinations. Basis of this model is a ray-tracing software used in conjunction with an optimization logic to determine the optimal component combinations. The details will be discussed in section 4.2 because the concept will also be integrated into a tolerance analysis tool.

Major difference to classical compensation (e.g. alignment) is the pre-determination of optimal configurations based on evaluations of the model. In addition, application of combinatorial assembly is restricted to a set of systems as combinatorial assembly requires different options to choose from. While theoretically every component parameter can be characterized and used, it will be more practical and less costly to use only a few of the available components. The number of relevant parameters per component plays a key role, as only components can be exchanged for each other. This makes it highly desirable to use only few significant parameters per component.

Being an alternative compensation strategy, combinatorial assembly has multiple uses for optical systems. It can be employed to increase the optical performance compared to classical assembly avoiding other compensation techniques such as alignment. Whereas common compensation strategies are almost exclusively iterative and require adjustable mounts, combinatorial assembly requires neither. Hence, through combinatorial assembly, assemblies that require compensation can be reduced in size and their mechanical and thermal robustness increased. In addition, parameters that can otherwise not be employed for compensation (such as material properties and component attributes like curvatures) may be used as compensators. A large number of compensators can thus be used, enabling simultaneous correction of multiple errors. It is this point that makes combinatorial assembly applicable to a wide range of optical design problems circumventing the drawbacks selective assembly suffers from, particularly mismatch and lack of multi-dimensional compensation for small volume production.

3.1.2 The fundamental matching problem

In Figure 15 it was illustrated that some component combinations are better than others. Finding the best component combinations is the heart of combinatorial assembly and classifies as a matching problem. Matching problems are a class of assignment problems that are common in operations research and other sciences [SRIN07]. This section defines the fundamental matching problem for optical systems and gives a mathematical description.

An optical system consists of optical and mechanical components (e.g. lenses and mounts) as was shown for the laser focusing lens in Figure 14. Every component has a number of attributes – or parameters – each subject to tolerance deviations given by a specific tolerance distribution. Figure 18 exemplarily shows modules and parameters of a laser focusing lens.

Figure 18
Schematic structure
of an exemplary
system for assembly

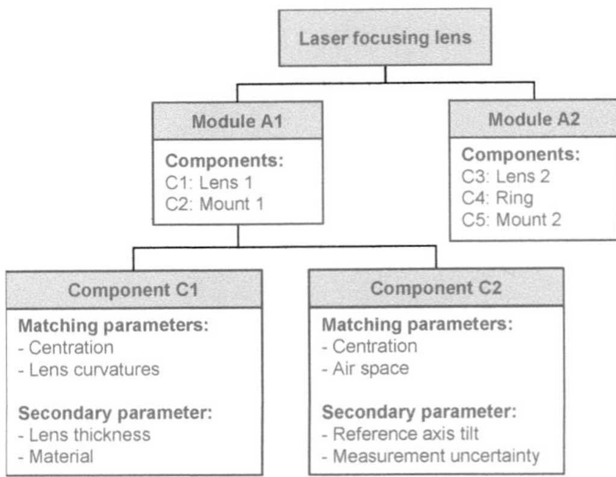

It will be assumed, that a given quantity n_i of every component type is available for production. Thus, during the assembly of a series of n systems, multiple components of the same component type are available, each with a number of attributes subject to tolerance deviations as shown in Figure 18.

To simplify the discussion, the following definitions will be introduced, mainly to distinguish real and abstract objects:

Optical system design – the layout or concept of a system

Optical system – a physically tangible assembly of components

Component type – the elements found in an optical system design

Components – physical components found in an optical system

Module – a single component or subassembly of components

Attribute or parameter – a characteristic feature of a component type

Parameter value –value of a component's attribute

Matching parameter – an attribute that is measured and used for selection

Secondary parameter – an attribute that is not measured (costly or insensitive)

During the assembly of a series of optical systems of the same design, combinatorial assembly creates a unique relation between available components. For the two-dimensional example in Figure 15 this was easily shown. Graph theory, a mathematical discipline used to describe relations between objects, is well suited to illustrate matching for higher dimensions. Figure 19 exemplarily depicts a matching graph for a system consisting of three components with a total of n systems being assembled.

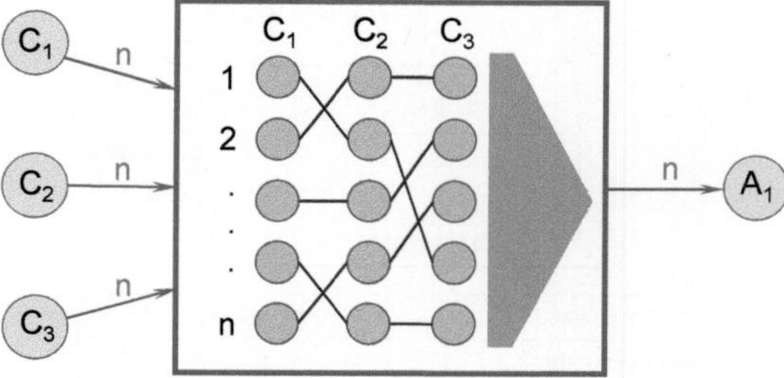

The graph consists of nodes, representing components C and connections between them that constitute a path. The second component of module C_1 is for example assembled with the first components of modules C_2 and C_3. Obviously, a system is completely assembled if the path connects exactly one component of every type.

In this example, every component is associated with a unique system and no components are left over. This constitutes a maximum matching problem [SRIN07] and we shall call this the fundamental matching. For better understanding, components of a specific type are arranged in columns. Connections between nodes of a column of the same component type are not allowed rendering the graph multipartite. The resulting output shall be called a set of systems and the selection can be mathematically described by a selection matrix M that defines the mapping between components:

3.1

$$M = \begin{pmatrix} C_{1,1} & C_{1,2} & \cdots & C_{1,m} \\ C_{2,1} & \cdot & \cdot & C_{2,m} \\ \cdot & \cdot & \cdot & \cdot \\ \cdot & \cdot & \cdot & \cdot \\ C_{n,1} & \cdot & \cdots & C_{n,m} \end{pmatrix} = \begin{pmatrix} 1 & 2 & 4 \\ 2 & 1 & 1 \\ 3 & 3 & 2 \\ 4 & 5 & 5 \\ 5 & 4 & 3 \end{pmatrix}$$

Every row of the selection matrix uniquely defines one system and gives the indices of the combined components. Every column is thus a permutation of the vector going from one to n. As the order of possible combinations does not matter, one of the columns can be freely chosen as a reference (e.g. the first column in increasing order). The total number of unique possibilities – or graphs – to combine n components of m different types hence calculates as the product of m-1 permutations and amounts to:

3.2

$$N_{comb} = n!^{(m-1)}$$

Even for the example in Figure 19 with very few components a total of $5!^2$ or 14400 possibilities exist. This number of possibilities needs to be distinguished from the number of individual system combinations which is the number of individually different systems. For these only n^m possibilities exist. The number of possible combinations in a set of systems is consequently larger, as a single system can be part of different sets of systems (the $n!^{(m-1)}$ possible permutation combinations).

Combinatorial assembly aims at finding the graph delivering a set of n systems which performs best. In order to find the optimal matching the performance of an entire set of n systems needs to be assessed. This is very important and conceptually different from other assembly strategies where every single system is optimized. In general, the performance of a set of systems depends on the selected systems, while the performance of an individual system depends on the selected components. The optimization problem can be formulated as finding the selection matrix M of an allowable selection $\mathbf{M}$ which maximizes the performance Q of a set.

3.3

$$\max\{Q(M)\,|\,M \in \boldsymbol{M}\}$$

Solving the optimization requires methods to determine the performance of a selection and will be discussed in section 3.2.

3.1.3 Variants of the matching

In addition to the fundamental matching just presented, combinatorial assembly can be performed in different variations. Differentiations of the variants can be made according to the number n of available components of a particular type, the number of component types m and the optimization criterion.

The number of available components does not need to be equal for every component type. That is n_i is not equal to n_j. If $n_i > n_j$, a maximum matching can only be performed for n_j systems and inevitably some components of type i will be left over (see Figure 20). A special case of this kind results if $n_j = 1$.

Figure 20
Graphs of matching variants:
incomplete matching (left), component symmetry (center), system symmetry (right)

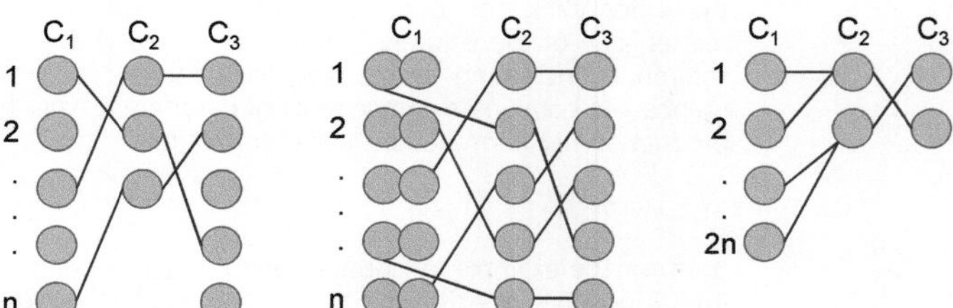

Instead of pursuing a maximum matching, scrap can be allowed regardless of the equality of the n_i. This intentional reduction of the number of produced systems can increase the performance of the remaining set even further. In a way, the worst systems in the set are just left away. An alternative optimization criterion for the matching could then be to maximize the number n of systems within required quality limits q_{min}.

3.4

$$\max\{n\,|\,q_i \geq q_{\min}\}$$

Any kind of weighting of excess parts is also possible. However, this work and the following discussion focus on the optimization of a complete set. Minimization of scrap will not be discussed.

Figure 20 (center and right) illustrates symmetrical designs. Symmetries are common properties of optical systems. For ease of manufacture, components are sometimes designed to have the same curvature on both sides and frequently the same type of component is used more than once in an optical system design. This has already been the case for the example in Figure 14. In addition, optical surfaces often exhibit symmetries and components can therefore be assembled in multiple configurations (e.g front-back symmetry). In the extreme, a rotationally symmetric lens can be turned and thus has an infinite number of possible configurations. While in this case regular compensation is possible, component orientation might be limited to discrete positions (e.g. cylindrical lens) and therefore require a combinatorial assembly approach. Owing to tolerance deviations, the symmetry that exists in an optical system design disappears in reality. The two curvatures of a biconvex lens will be slightly different and one lens is never exactly like the other. Thus with tolerances, orientation matters and requires the evaluation of all possible arrangements. Modeling symmetries with graph theory can be achieved by adding possible assembly states and imposing restrictions to the allowable relations between objects as shown in Figure 20.

Whereas symmetrical components call for split components where only one can be used (Figure 20, center), multiple use of the same component type (system symmetry) can be described by multiple paths connecting to one component type (Figure 20, right). This reduces the number of possible partitions and increases the number of possibilities for one symmetry to:

3.5
$$N_{Comb} = \frac{(k \cdot n)!}{k} \cdot n!^{\,(m-k)}$$

Here, m and n are the number of component types and the number of components as before and k is the number of symmetry states. Hence, if symmetry would have been considered in the example, the number of possibilities rises to 10!5!=217e6.

3.2 Optimization of component matching

Finding the optimal combination of components to maximize the performance of optical systems is the core of the matching problem. It is not only important during real assembly, but also during conception and tolerancing.

Let q_i denote again the performance of a single system. As multiple systems of the same type are assembled during one production run their combined performance needs to be optimized in order to increase the cumulated probability $\Phi_{cum}(q)$. Hence, a figure of merit, representing the performance Q of such a set of systems needs to be found. Some candidates for a performance measure are:

- the arithmetic average performance
- the root sum square of individual system performance
- the worst individual system performance

Typically, not the performance q of a system but the performance error is measured and shall be denoted $q*$ (e.g. merit function). The above measures will thus translate to:

$$\text{3.6} \qquad Q_1^* = \frac{1}{n}\sum_{i=1}^{n} q_i^* \qquad\qquad Q_2^* = \frac{1}{n}\sqrt{\sum_{i=1}^{n} q_i^{*2}} \qquad\qquad Q_3^* = max\{q_i^*\}$$

All of these measures can be useful and even more sophisticated measures are possible. The choice is application driven. Optimization aims at the reduction of the quality measure $Q*$ as it measures the deviation from optimal quality. The general optimization problem of finding a selection M thus reads as:

$$\text{3.7} \qquad min\{Q^* \mid M \in \mathbf{M}\}$$

Optimizing for arithmetic average performance is likely to produce excellent systems as well as systems of minor quality, while the root sum square will minimize the variance of the performance error without taking the mean into consideration. Taking the worst individual performance as the optimization measure can be very useful for small sets of systems, raising the performance over a critical threshold. Figure 21 shows histograms of q_i^* for sets of n systems, highlighting the differences of the three performance measures Q_i* mentioned above. The curves are simulated distributions for a linear combination model with two random variables and Gaussian probability distribution.

Figure 21
Performance distributions of a system with linear performance function for optimized quality measures Q

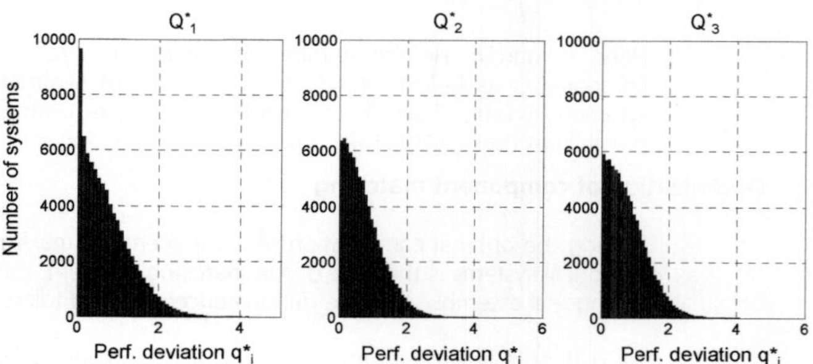

To summarize, the quality of a series production is measured as a single value Q and optimization of this measure yields the optimal matching configuration. The next sections will discuss the optimization.

3.2.1 Rigorous calculation

Rigorous calculation implies testing every possibility. Even though calculating every single component combination and selecting the best cannot be called optimization, it is the only method that surely delivers the global optimum. It will thus serve as a benchmark, both in terms of computational effort and optimality of the solution and is well suited to introduce the problem of optimizing the selection.

As demonstrated in chapter 3.1.2 the total number of possible selections amounts to $n!^{(m-1)}$ (see equation 3.2) with n being the number of systems and m the number of dimensions, that is the number of component types in a system. The optimization procedure can be performed in two different ways. It is possible to iterate through every of the $n!^{(m-1)}$ component selections M and determine the quality Q of the selection. In practice, however, not every selection of components is entirely different from another as the total number of different component combinations is n^m. The alternative analysis procedure makes use of this fact and determines the individual q_i or q_i^* first, while the selection of the optimal set can be performed separately. Then, for every selection M the quality criterion Q can determined (Figure 22, right) and the optimal one chosen.

Figure 22
Nassi Shneiderman
diagrams of
rigorous analysis
alternatives

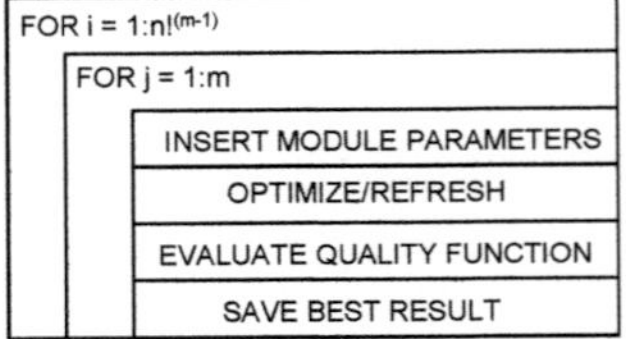

For both procedures the measured parameter values are inserted in the model and then additional compensators (e.g. back focus) adjusted. As the effort to determine the individual q_i requires ray tracing operations, evaluation of arbitrary performance criteria and often optimization of additional compensators, computation time should be kept low. A comparison of the two strategies shows a clear advantage for the two step approach if the time t_{eval} required to evaluate a lens is much longer than the time t_{comp} to compare the numerical results, and if n is large (see 3.8).

3.8
$$t_a = n \cdot n!^{m-1} \cdot t_{eval}$$

$$t_b = n^m \cdot t_{eval} + n \cdot n!^{m-1} \cdot t_{comp}$$

A simple procedure for the calculation of all n^m possible combinations (e.g. the first loop of the two step approach) consists of cascaded loops for every component. For practical purposes this is not very useful as the number of component types m in a set will be different for every application and so would be the number of loops. Rearranging the problem allows to perform the calculation for arbitrary numbers of n and m with only four cascaded loops. Figure 23 compares the two methods.

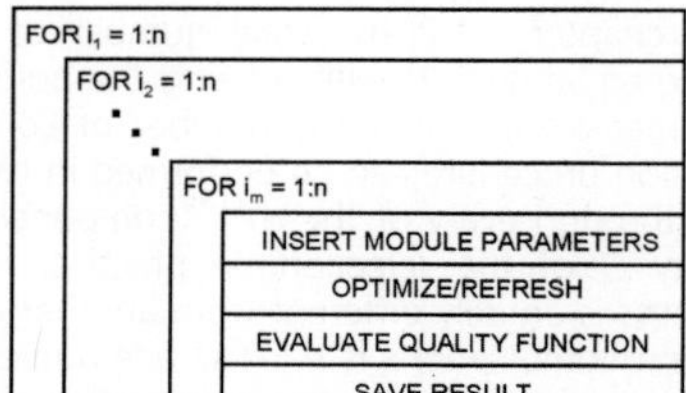

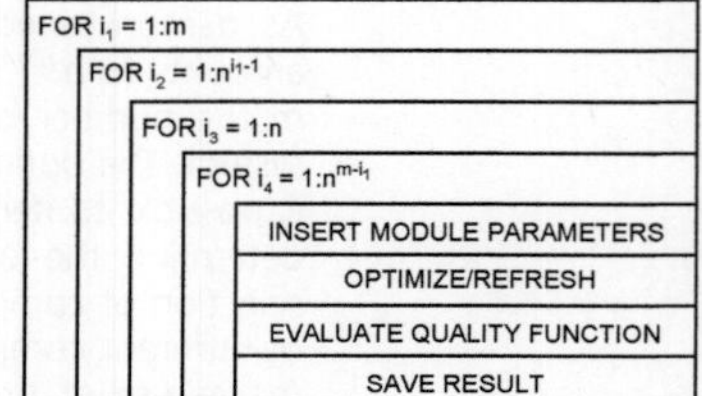

Figure 23
Nassi Shneiderman diagrams of cascaded loops (left) and transformed description (right)

Transforming the problem to four loops is achieved by recognizing that a list of combinations is ordered in blocks of components of the same type as illustrated in Figure 24. The first loop selects the module while the second determines the number of blocks. The third loop selects the number while the fourth determines the number of repetitions for every number.

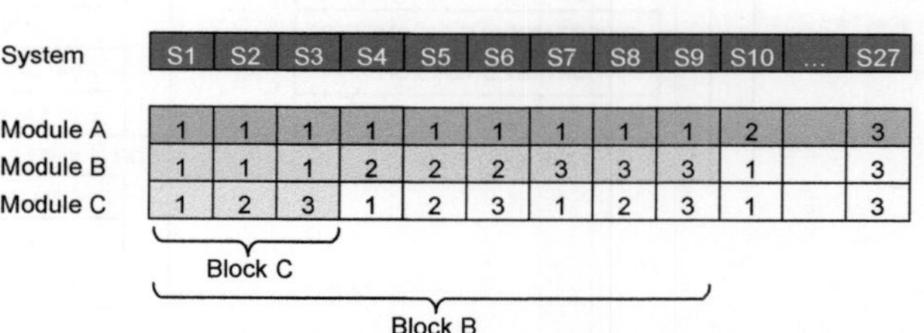

System	S1	S2	S3	S4	S5	S6	S7	S8	S9	S10	...	S27
Module A	1	1	1	1	1	1	1	1	1	2		3
Module B	1	1	1	2	2	2	3	3	3	1		3
Module C	1	2	3	1	2	3	1	2	3	1		3

Figure 24
Calculation of all component combinations

In this example block C contains numbers from one to n and is repeated n^{m-1} times. For block B every number from one to n is repeated n times.

In order to determine the $n!^{(m-1)}$ different selections during the second part of rigorous calculation, it is useful to remember that every column of the selection matrix is a permutation (compare chapter 3.1.2). It is then possible to arbitrarily choose the first permutation as the order of systems is arbitrary, and successively loop through the permutations of the other components. For the generation of permutations standard algorithms exist and can be used. A lexicographic order is most suitable in particular for the optimization that will be discussed subsequently.

A comparison of the rigorous calculation for different lenses and number of systems shows that computation time t_{eval} is in the order of milliseconds (typical 100 ms) such that evaluation of five systems (n) and three modules (m) takes approximately 12.5 seconds for the considered example. Such a duration is fine for a single optimization, but during tolerance analysis the calculation needs to be repeated many times.

3.2.2 Linear optimization

Instead of testing every possible combination the selection can be found through optimization. Matching problems require discrete optimization, either linear or nonlinear depending on the formulation of the problem. Similar to the rigorous calculation, two general optimization strategies are possible:

1. an optimization based on n^m predetermined quality values of component combinations

2. an optimization including the evaluation of a combination's quality

Predetermined quality values are used during rigorous calculation due to their advantage in terms of computation time. For optimization the type of the optimization problem, its difficulty and quality of the solution play very important roles, too and also depend on this choice. If no evaluation of the different component combinations is performed in advance the optimization task necessarily becomes highly nonlinear. The possible solution is composed of quality values of randomly scattered component combinations. This could be minimized if modules are brought into a sorted order. Figure 25 shows a two dimensional parameter space (two modules with one parameter each) to illustrate the nonlinear optimization. In every optimization step the performance measure of the selection M is determined. Hence, same configurations, represented as dark dots, are evaluated multiple times.

Figure 25
Graphical illustration of nonlinear optimization of combinatorial assembly selection

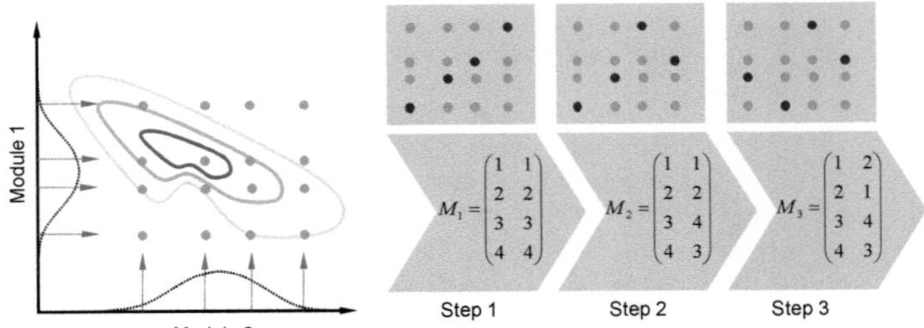

Predetermined values, on the other hand, require the evaluation of particular component combinations that – in the end – will never be used. However, depending on the performance measure Q used to characterize the optimal selection the optimization can be expressed as a binary integer problem even though the performance function landscape can be highly nonlinear. In addition, the number of predetermined values is small compared to the number of permutation possibilities. The advantage over non-linear optimization is the fact that local optima are not an issue. The problem is reduced to finding the optimal permutations and allows formulation of the optimization problem in the standard form of linear programming [VAND08]:

3.9
$$\min\{Q^* = q^{*T} x \mid Ax = b\}$$

As mentioned before, the task of finding optimal matches falls into the category of assignment problems, an important class of optimization tasks especially in the operations research. In order to bring the multidimensional matching problem to the standard formulation of linear programming the evaluated performances of every possible component combination must be brought into vector form in an ordered fashion. The optimization process then reduces to a binary integer program defining the selection of a set of systems. Elements of the selection vector x can be either one or zero (e.g. n elements are equal to 1) and boundary conditions must be formulated to guarantee that only physically feasible selections are chosen. Of the above mentioned measures, both, the average quality and the root sum square measure lead to a linear formulation of the optimization problem.

3.10
$$Q_1^* = \frac{1}{n} \cdot \begin{pmatrix} q_1 \\ .. \\ .. \\ q_{n^m} \end{pmatrix}^T \cdot \begin{pmatrix} x_1 \\ .. \\ .. \\ x_{n^m} \end{pmatrix} \qquad Q_2^* = \frac{1}{n^2} \cdot \begin{pmatrix} q_1^2 \\ .. \\ .. \\ q^2{}_{n^m} \end{pmatrix}^T \cdot \begin{pmatrix} x_1 \\ .. \\ .. \\ x_{n^m} \end{pmatrix}$$

While this is fairly obvious for the average quality measure it might be striking at first for the root sum square measure. However, the minimum of the root sum square is equal to the minimum of the sum of the squares and the optimization task reduces to a linear combination of squared values.

An important step during linear optimization is the definition of boundary conditions to limit the optimization to valid results. Matching requires every component to be used only once. In the case of a binary integer program with the formulation as stated above, boundary conditions take the following form:

3.11
$$A = \begin{pmatrix} 1 & 1 & 0 & 0 \\ 0 & 0 & 1 & 1 \\ 1 & 0 & 1 & 0 \\ 0 & 1 & 0 & 1 \end{pmatrix}, \quad b = \begin{pmatrix} 1 \\ 1 \\ 1 \\ 1 \end{pmatrix}, \quad q = \begin{pmatrix} q_{1,1} \\ q_{1,2} \\ q_{2,1} \\ q_{2,2} \end{pmatrix}$$

If the q_i are ordered in lexicographic order as depicted in Figure 24 the A_{ij} can be analogously defined to be either one or zero. As the definition of the boundary conditions changes with the number of components and the number of modules it is necessary to automatically generate the boundary conditions. Figure 26 illustrates the boundary conditions for an example with 3x3 systems and modules.

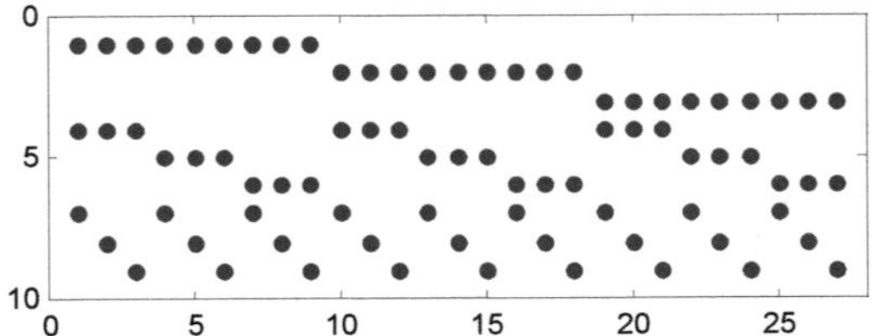

Symmetrical problems demand additional boundary conditions that depend on the structure of the value vector. For the two separate cases of module and system symmetry Figure 27 illustrates the situation. For symmetrical components the number of possibilities for that component is multiplied. For symmetrical modules every module is multiplied. The boundary conditions then need to guarantee feasibility by restricting the selection. In the examples this can be achieved by restricting the sum of the highlighted rows and columns such that only one selection is made. Note that in the example (right) the sum of rows and columns is considered, as four modules of the same type are present and distributed to two locations in the system. A feasible selection is highlighted.

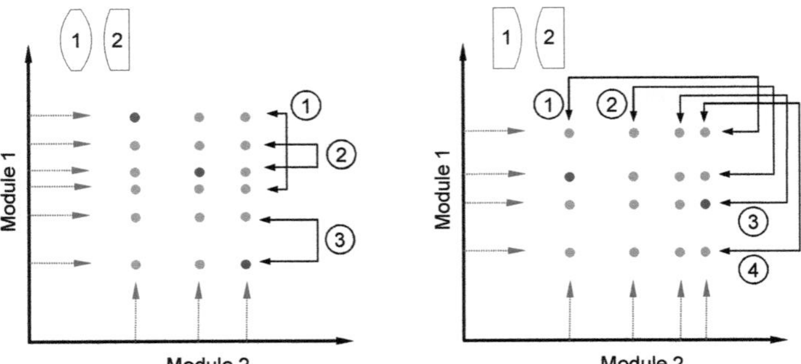

Solving the optimization problem requires algorithms such as Branch and Bound that can find optima for a binary integer programming problem [VAND08]. Execution time depends on the starting point as well as on the size of the problem. An estimate of the computation time t_{opt} therefore contains an average time t_{aver} of the optimization that depends on the number of systems.

3.12

$$t_{opt} = n^{m} \cdot t_{eval} + t_{aver}(n)$$

Figure 28 shows a comparison of the time required to determine the optimal selection using rigorous calculation and optimization illustrating the dramatic increase in time of the rigorous method, once a critical number of systems is reached.

Figure 28
Execution times of
optimization
strategies

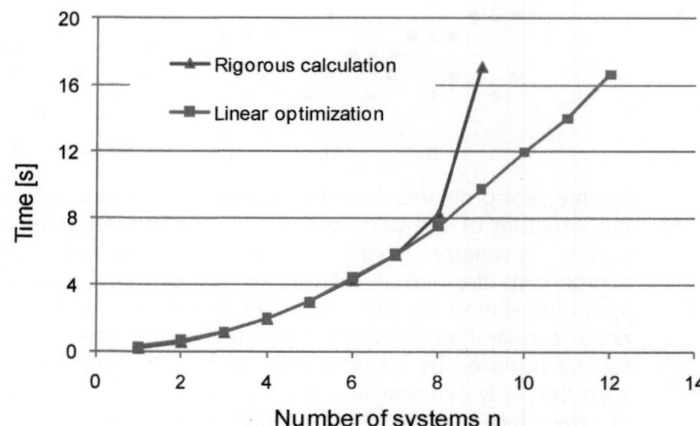

The effort during optimization is largely determined by the calculation of component combinations, while the optimization itself terminates quickly. This can be enhanced if approximations are used to limit the number of combinations that are determined.

3.2.3 Approximate optimization methods

Whereas rigorous calculation and optimization can lead to the optimal solution, both methods can be very time consuming for larger problems. Approximations and heuristic methods can help to reduce computation time. Heuristics are strategies for finding solutions to a problem with limited time and knowledge [ROTH11]. Applied to the matching problem, heuristic procedures can locate or approximate optima based on a set of rules. A number of heuristics exist for the solution of linear programming tasks and they are employed when optimization algorithms fail, are too complex or slow. The Hungarian method for example solves weighted assignment problems on bipartite graphs [FRAN11]. This is already very similar to the problem at hand. So called Greedy algorithms search for the next optimal steps and crawl their way forward [FRAN11].

Under the simplifying assumptions of a linear quality function of a system and an arithmetic average performance measure Q^{*}, of a set of assembled systems a very interesting and simple solution is obtained. This shall be illustrated for a two dimensional problem. Assume a linear quality function q of a system (e.g. the quality function is a linear combination of the parameters x_i with sensitivities S_i.

3.13

$$q = \left| S_1 x_1 + S_2 x_2 \right|$$

For a constant quality level the parameter values lie on a straight line in the x_1-x_2 space. In this case the arithmetic average quality Q^*, of a set of systems with quality q is optimized if components are sorted according to their parameter values and then matched (Figure 29).

Figure 29
Linear quality
function
optimization

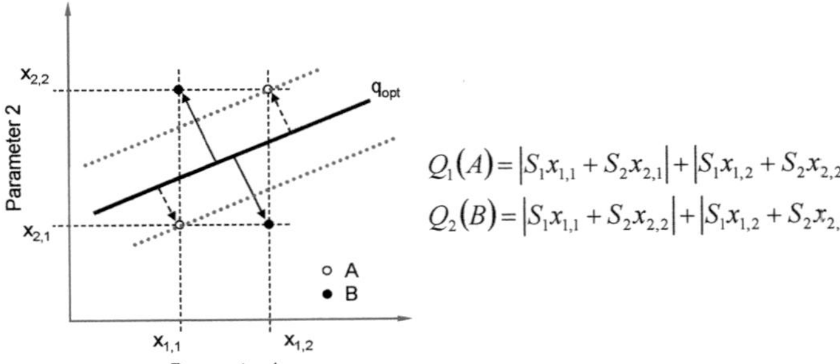

$$Q_1(A) = \left| S_1 x_{1,1} + S_2 x_{2,1} \right| + \left| S_1 x_{1,2} + S_2 x_{2,2} \right|$$

$$Q_2(B) = \left| S_1 x_{1,1} + S_2 x_{2,2} \right| + \left| S_1 x_{1,2} + S_2 x_{2,1} \right|$$

This is very intuitive and can be shown in Figure 29. For any two non-neighboring points in the diagram, the distance to the optimal solution is the sum of two line segments. As long as combinations lie on either side of the optimal line (e.g. terms in the formula are positive and negative), sorting will be optimal. In all other cases orientation does not matter at all and hence sorting is generally advisable. The optimal arrangement for four combinations is thus found and the principle can be extended through induction for the complete problem with a larger number of combinations. It is therefore possible to sort component parameters in increasing or decreasing order and combine components in the established order. From a sensitivity analysis of the quality function the suitable order (increasing/decreasing) can be determined for every attribute. While this works fine if every component has only one significant attribute the strategy needs to be adapted for components with multiple parameters as only one sorting can be carried out. In vector notion the problem can be expanded to multiple dimensions. As long as the problem can be exactly solved the reasoning holds true.

Compared to rigorous calculation, the described method only requires sorting of parameter values and the evaluation of n systems. The computational effort for sorting is comparably small. Even for slow algorithms sorting of several thousand values terminates in a fraction of a second. Hence, the required time

t_{heur} to perform the heuristic analysis is limited by the evaluation time t_{eval} of the pre-selected systems.

3.14
$$t_{heur} = n \cdot t_{eval}$$

For many applications a linear quality function is a very good approximation because tolerance deviations are small, but the criterion needs to be carefully selected. Even though an average quality is not generally desirable, a strong correlation of the different optimization criteria guarantees if not a perfect, a good solution.

In practice, it is not even necessary to carry out the sorting but use the approximate description of the performance q to determine the optimal matching configuration. It is then possible to determine first (and second) order derivatives and base the selection on approximation. To improve the solution, a substitute performance specification which can be linearly expressed can be used. As mentioned before, Zernike polynomials or aberration coefficients can serve this purpose and may be used to select optimal matching without the need to employ ray-tracing calculations. In the final step these pre-selected values need to be accurately evaluated with ray-tracing calculations to confirm optimality of the solution. If such a procedure is too inaccurate, the solution obtained can provide a good starting point for search algorithms and can reduce optimization time. It is then necessary to evaluate module combinations in the vicinity of the approximate solution and apply search algorithms to find the optimal solution of the reduced problem.

3.3   Application to optical systems

The purpose of combinatorial assembly is the cost-efficient production of high performance optical systems. So far, the principle of combinatorial assembly has been presented and optimization methods are given to find the best configuration of characterized components. In order to successfully employ combinatorial assembly it is not sufficient to arbitrarily measure component parameters and select the best combination. Instead, the application requires a careful choice of matching parameters and suitable components or subassemblies during design.

In addition, the performance function and the optimization criterion need to be defined and the number n of a production run determined. Furthermore, lens design and tolerance allocation can be changed and a number of other boundary conditions need to be satisfied. Figure 30 provides an overview of the key elements for optical systems.

Variables	Constraints	Specifications
• Lens design • Assembly sequence • Number and structure of modules • Production volume • Tolerance limits	• Possible modules • Available metrology • Manufacturing precision limits • Mechanical boundary conditions	• Basic system properties • Imaging performance • Cost of manufacture • Cost of tolerances • Cost of assembly processes

For the following discussion it will be assumed, that a given lens design is analyzed and strategies will be developed to determine a suitable set of matching parameters, modules and tolerances. The discussion of design changes is left to chapter 5.

3.3.1 Combinatorial compensation of basic system properties

Despite the intuitive character, there is no reason that combinatorial assembly will always produce systems of higher quality. The fundamental precondition is the existence of a quality function that is suitable for compensation.

Compensation can be explained on different levels if the performance degradation T due to tolerances is expressed as a power series expansion. As basic system properties such as focal length and magnification are paraxial quantities, derived by substituting the sine in Snell's law by its argument, the resulting relations are linear. The performance degradation T for a single specification can thus be expressed as follows with $S_i^{(1)}$ a vector containing the sensitivities:

3.15
$$T = \underline{\Delta x} \underline{S}^{(1)} = \sum_i \underline{c}_i \underline{S}_i^{(1)}$$

For combinatorial assembly, parameters cannot be changed entirely independently due to the discrete nature of modules. Hence, individual components C_i are represented by a vector $\underline{c}_i$ containing the changes Δx of parameter values x from their nominal values such that only the parameters values of that particular component are allowed to be non-zero. Ideally, the effect of a perturbation induced by one component is completely compensated during combinatorial assembly such that T equals zero.

In the simplest case of a single specification a linear equation results from equation 3.15. This is shown for Zernike coefficient three (wavefront tilt) of the laser focusing lens example from Figure 14 for modules A_1 and A_2 with one and two relevant parameters respectively. The first module contains lens one, while the second contains lenses two and three. Relevant parameters in this example are the three lens centrations, here denoted as x_1, x_2 and x_3 and units are inverse mm.

3.16

$$T = \begin{pmatrix} S_1 \\ S_2 \\ S_3 \end{pmatrix}^T \cdot \begin{pmatrix} \Delta x_1 \\ \Delta x_2 \\ \Delta x_3 \end{pmatrix} + \begin{pmatrix} S_1 \\ S_2 \\ S_3 \end{pmatrix}^T \cdot \begin{pmatrix} \Delta x_1 \\ \Delta x_2 \\ \Delta x_3 \end{pmatrix} = \begin{pmatrix} 14.8 \\ -8.72 \\ -6.68 \end{pmatrix}^T \cdot \begin{pmatrix} \Delta x_1 \\ \Delta x_2 \\ 0 \end{pmatrix} + \begin{pmatrix} 14.8 \\ -8.72 \\ -6.68 \end{pmatrix}^T \cdot \begin{pmatrix} 0 \\ 0 \\ \Delta x_3 \end{pmatrix}$$

The performance change induced by deviations of one component can then be compensated by parameter changes of the other component. Compensation of a single specification is hence possible with only two modules as they provide enough degrees of freedom.

In addition, the tolerance ranges need to match such that the effect of a deviated component is never larger than the combined effects of all other components. This maximizes the compensation because otherwise perfect compensation is not always possible. For a single specification and two modules, the two terms in equation 3.16 need to be equal if tolerances are at their limits t. This is equivalent to a worst case criterion, alternatively a RSS summation can be used if tolerances are assumed to be Gaussian.

3.17

$$\begin{pmatrix} |S_1| \\ |S_2| \\ |S_3| \end{pmatrix}^T \cdot \begin{pmatrix} t_1 \\ t_2 \\ 0 \end{pmatrix} = \begin{pmatrix} |S_1| \\ |S_2| \\ |S_3| \end{pmatrix}^T \cdot \begin{pmatrix} 0 \\ 0 \\ t_3 \end{pmatrix}$$

For more than one specification the compensation situation gets more complicated. Instead of a sensitivity vector, a matrix of sensitivities for the different specifications needs to be employed. The developer is left with a set of linear equations and variables (modules) that in general impact multiple specifications. The possible degree of compensation then clearly depends on the number of available degrees of freedom as well as on the interrelations of specifications. Equation 3.18 shows the situation for the compensation of two basic system properties of a system with two modules. Module one consists of parameters x_1 and x_2 and module two contains parameter x_3. $S_{j,i}$ are the sensitivities of specification j and parameter i.

3.18

$$T = \begin{pmatrix} S_{1,1} & S_{1,2} & S_{1,3} \\ S_{2,1} & S_{2,2} & S_{2,3} \end{pmatrix} \cdot \begin{pmatrix} \Delta x_1 \\ \Delta x_2 \\ 0 \end{pmatrix} + \begin{pmatrix} S_{1,1} & S_{1,2} & S_{1,3} \\ S_{2,1} & S_{2,2} & S_{2,3} \end{pmatrix} \cdot \begin{pmatrix} 0 \\ 0 \\ \Delta x_3 \end{pmatrix}$$

Full compensation can be achieved if the effects of random perturbations of one module can be completely reduced by the remaining modules. In the above example, two equations need to be satisfied and module two has not enough degrees of freedom to satisfy both. Hence, full compensation is not possible. However, a limited compensation is still possible but the probability of reducing an error is smaller than for just a single condition. This is illustrated in Figure 31 where two specifications are shown together with a perfect compensation of deviations from component one and two.

Figure 31
Three dimensional
compensation

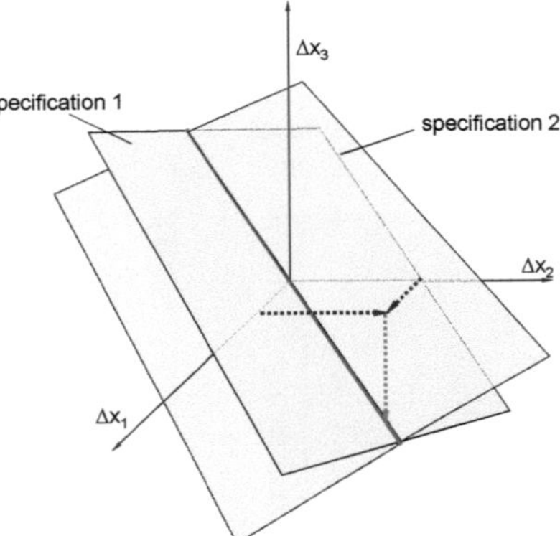

Perfect compensation is only possible at the intersection of specifications. In the three dimensional case this is a line, while for higher dimensions the intersections represent hyper planes. For any combination of modules the goal is to get as close as possible to this line. If correlations between specifications are strong it is not as important to meet the exact intersection of specifications but to get close to it. The problem then shows a reduced number of dimensions. This is comparable to the rank of the set of linear equations (or the corresponding matrix).

If additional modules are used, the beneficial compensation effect can be increased. Two mechanisms are of relevance: first, the fraction of the total tolerance error that can be compensated for is larger because more parameters are characterized, and second, the number of possible combinations increases enabling better compensation. A tradeoff between cost and performance increase may be found to guarantee cost-efficiency.

3.3.2 Compensation of imaging performance criteria

In contrast to basic system properties, Optical imaging specifications will not generally follow a linear behavior. They are usually given as wavefront errors, resolution or other ray based measures and the calculations of these quantities require a large number of rays to be traced through the lens. Despite the complexity, a performance change induced by a parameter tolerance can often be compensated by a proper adjustment of other parameters as can be experienced during final tune up of lens design when parameters are slightly altered.

Imaging performance is in general evaluated for multiple points across the field of an optical system and also evaluated at multiple wavelengths. If specifications for every field point, wavelength or configuration are given, many criteria need to be satisfied making combined measures such as the merit function very appealing. An analysis of the relations between parameter changes and combined performance measures typically show a behavior as depicted in Figure 32.

Figure 32
Two-dimensional tolerance region of a merit function containing multiple criteria (image quality and focal length)

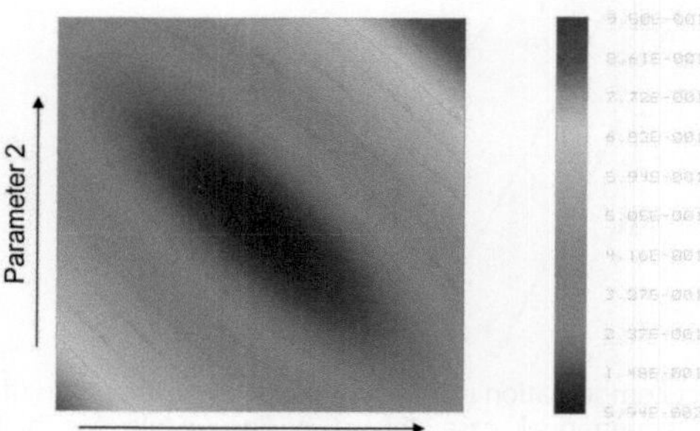

Highest performance is achieved in the central optimum and the change due to a single parameter perturbation is often nonlinear for very small changes and asymptotically reaches a linear behavior. For two and more parameters many performance criteria show stretched valleys or elliptical regions with areas of almost constant performance. Similar depictions of these valleys can be found in [GROS07] and are thought to be a common characteristic of optical systems. It is due to the nature of combined measures and the combination of field points, wavelengths and configurations that the performance exhibits a nonlinear behavior.

In order to represent this in the model, higher order sensitivities can be included. If $S_i^{(1)}$ is the vector containing the sensitivities, $S_i^{(2)}$ a matrix containing the second order sensitivities and c_i the component vectors, the following results:

3.19
$$T = \sum_i \underline{c}_i \underline{S}_i^{(1)} + \underline{c}_i \underline{\underline{S}}_i^{(2)} \underline{c}_i^T$$

Through these higher order errors the equations describing compensation become nonlinear further complicating compensation. The number of equations, corresponding to the number of specifications, however stays the same. But the degree of compensation will now reduce the compensation effect as the nominal performance can never be reached through compensation.

Measuring and identifying compensation with a nonlinear behavior is a difficult task, in particular for a large number of tolerances. It is therefore useful to search for alternative representations avoiding the particularities of the merit function. The most promising candidates are Zernike polynomials, describing wavefront changes, and aberration theory.

Zernike polynomials decompose the wavefront of a field point into orthogonal aberrations. Instead of correcting a physical measure such as the RMS wavefront error or Strehl ratio, compensation for the individual aberrations can be targeted. It is thus necessary to break the performance degradation down to specific aberrations and determine the effect of parameter perturbations on the Zernike coefficients. As they behave approximately linearly, this is easily done with a sensitivity analysis. The linear model of compensation can then be employed and one specification for every critical coefficient created.

As an alternative, aberration theories can be used. They are based on an approximation of the laws of refraction and reflection instead of approximations of the wavefront. Even though third- and even fifth-order theory are not sufficient to accurately predict the wavefront of a design the theories can be employed to predict the *changes* of a wavefront due to small parameter perturbations. The perturbed wavefront W_{Tol} can be expressed as a sum of the accurately determined wavefront W_{Nom} of the nominal system and the change W_{Aber} predicted by aberration theory.

3.20
$$W_{Tol} = W_{Nom} + W_{Aber}$$

If the nominal system performance is assumed to be the ideal state, compensation would have the goal to return to this origin. Consequently, if the approximation is reduced to a third-order (Seidel) level all the Seidel coefficients contributing to the performance measure should return to their initial values. For a centered monochromatic design it would thus be required to satisfy five constraints. Namely, changing spherical aberration, coma, astigmatism, field curvature and distortion back to their nominal values. Similarly, vector aberrations can be employed to set up specifications for decentered systems or those including chromatic aberrations.

In practice, the problem can be reduced if only one or two aberrations are of relevance. While higher order aberrations can be considered as well using Seidel's theory can give good approximations for small parameter changes. Its use is also justified by the observation that higher order aberrations tend to change slowly with parameter changes. Several advantages are gained: Aberration coefficients are easily calculated and defocus does not need to be included as a separate compensator because the Seidel coefficients are independent of defocus.

In using Zernike polynomials or aberration theory it is hence possible to reduce the understanding of image performance compensation to simple equations that behave almost linearly for small perturbations. It is even possible to interpret the behavior in Figure 32 as an overconstrained set of approximately linear functions. There is however a distinction to the linear models of basic system properties. While for basic system properties a clear specification is given for a single linear equation, the specification that aberration coefficients need to return to their initial values is artificially created. It is conceivable that a different balance of aberrations may perform equally well and hence this is not covered in the model. In addition, it might not be clear which aberration is most important during compensation having the largest effect on the original specification (e.g. MTF or Strehl-ratio). In this case, weights can be given to the different aberrations but their determination is subject to experience and trial and error.

If it comes to the question which model is more suitable for the application of combinatorial assembly, the answer depends on the problem. If compensation is analyzed for a given design that shall not be modified, Zernike polynomials are a good choice. If however design changes are considered, aberration theory can provide insight into the possible mechanisms and help the designer to optimize the design for compensation. The applicability of a particular model needs to be individually verified for every case and it may be necessary to include higher order errors to accurately predict performance compensation. For the purpose of understanding the principles of combinatorial compensation however, the models are very useful.

3.3.3 Mechanical realization of modules

Once a linear model of combinatorial assembly is derived, insensitive parameters can be excluded and declared secondary parameters before suitable modules can be defined which allow compensation. Additional graphical illustration of the magnitudes of sensitivities is very helpful to judge possible compensators and identify worst offenders as well as parameters of minor importance and possible correlations between specifications.

The task of selecting suitable modules reduces to grouping parameters to modules in order to solve the set of (linear) equations describing the compensation and finding appropriate tolerance limits. While some parameters will inevitably belong to a module (the parameters of a lens or lateral displacements of a mount) others can be more or less freely distributed to modules. During the process of defining moduls, it is advisable to examine the possibilities of the design at hand and tabulate feasible module definitions alongside with associated costs of measurements, mechanical design and potential cost.

Figure 33 shows alternative forms of lens mountings and compares their major benefits and drawbacks.

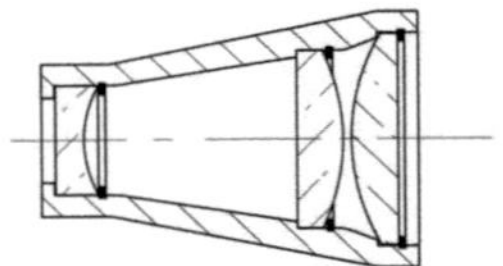

Drop-in assembly
+ Small number of parts
+ Compact and light
- Limited centration control

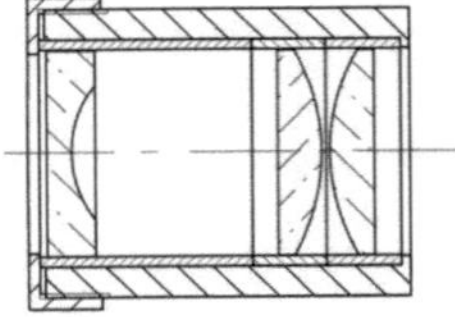

Poker chip
+ Standardized elements
+ Axial and lateral control possible
- Some lateral play necessary

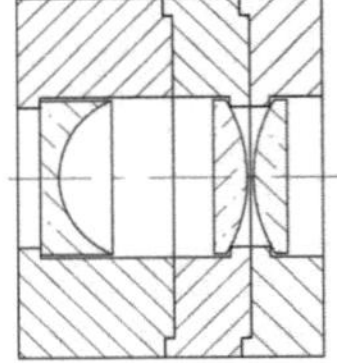

Modular assembly
+ Very high precision
+ Good lateral control
- High mechanical effort

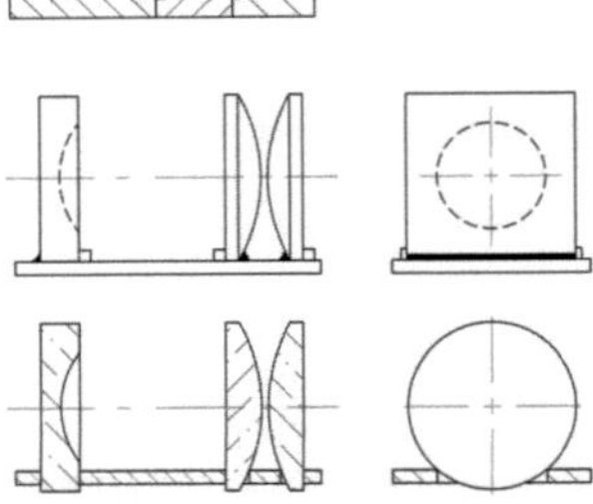

Planar Assembly
+ Suitable for robot assembly
+ Every element is a module
- Alignment restricted
- Potentially too many degrees of freedom

Drop-in assemblies, by the nature of this type of assembly, provide limited centration rendering this configuration mainly suitable for the compensation of symmetric errors that do not depend on centration. Focal length, distortion or spherical aberration are some examples. For alignment related errors however, a precise mechanical reference of modules is required to guarantee that modules will be in a predictable position. It is not necessary that mounts are manufactured to extreme precision, but measurements should enable precise predictions about the later connection of modules (e.g there shouldn't be any play). Mechanical setups such as "Poker chip" and modular assembly provide significantly better centration control and are thus suitable to compensate asymmetric errors arising from decentration and tilt in addition to symmetric errors. Special types of this form are lenses with integrated mounts made from plastic that can be realized for plastic lenses [TRIB64]. Alternative designs for anamorphic lenses or robot-based automated assembly can be based on planar setups as given in Figure 33. Centration compensation is equally possible with these configurations which may contain physical assists to control position or allow free positioning of components. Planar assembly has recently become more popular due to miniaturization and automation. Apart from these basic configurations combinations are often used. The exact design of mechanical mounts necessarily depends on the lens design, however, for small design changes which might be enough to optimize the application of combinatorial assembly the general type can often remain, as can the definition of modules.

In addition, it should be considered to use combined measures to realize the matching. For symmetrical specifications radii, thicknesses and glass properties represent constructional parameters that can be determined while focal length would be a combined measure. For asymmetric errors, centration of modules consisting of multiple elements can be determined. As tolerances of multiple parameters add up statistically, measures such as focal length can often be assumed to follow a Gaussian probability distribution according to the central limit theorem. This can be of great advantage as it reduces mismatch due to varying tolerance distributions. In terms of cost it can also be very profitable to measure a combined criterion as the number of individual measurement tasks is reduced. Custom measurement equipment to accurately determine the desired parameter values can provide very useful.

3.3.4 Cost-efficient application of combinatorial assembly

The costs of combinatorial assembly are determined by very different factors during design, manufacture and assembly such as

- structure and complexity of the lens design
- available manufacturing technologies
- assembly strategy and structure of modules
- tolerances of optical and mechanical components
- metrology
- equipment and personnel
- storage and selection of components

If the performance of an optical system is plotted over its cost of production, curves as shown in Figure 34 can generally be obtained, as tolerances (and other cost factors too) increase with performance [WILL83,DGRO09].

Figure 34
Cost-performance
relations for
standard and
precision assembly

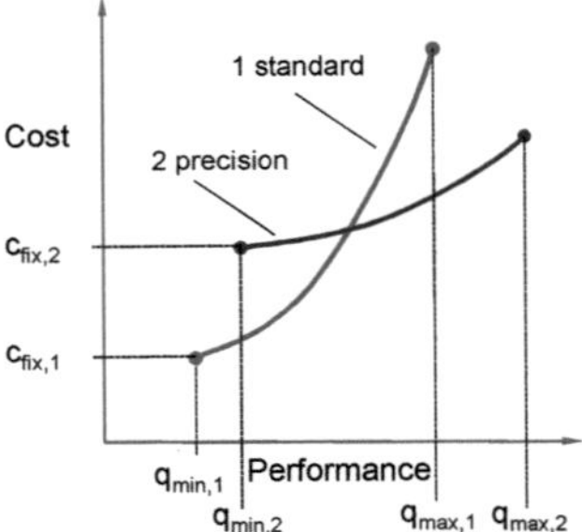

Ideally, a production with the minimum cost for a desired level of performance would be chosen during conception. This is of course a simplification because other, non-monetary, aspects may determine the success of a company. But for the purpose of this discussion only cost and performance will be considered. For high-performance systems, the goal is to select a strategy that lowers the marginal cost at high qualities and extends the maximum quality limit as far as possible ($q_{max,2} > q_{max,1}$). Hence, costs that depend significantly on performance need to be reduced primarily. Tolerance demands are thus very critical.

In order to compare different possible assembly strategies, it is important to have a figure measuring the success of a production. Creating a success function summarizing various goals is customary in decision oriented production theory [DYCK07] and in product development [PAHL06]. For the cost-efficient production of optical systems, this work defines a monetary success function Γ that is only based on cost C and performance q such that a production with higher values of Γ is regarded superior to a production with lower values.

3.21
$$\Gamma = \int \Phi(q) \cdot p(q)\,dq - C$$

With $\Phi(q)$ the probability of achieving a performance level q and p the price per unit, the success is given as the expected price per unit reduced by the cost C of manufacture and assembly. As both, the probability of achieving a performance level and the cost depend on manufacturing and assembly processes the success function can be quite complex, non-linear and discontinuous. The very general expression in equation (3.1) can be simplified if a) price is constant and b) the integral of $\Phi(q)dq$ is one (e.g. 100% yield is achieved). This special case is the typical situation assumed in scientific works on cost-optimal tolerancing (compare for example [YOUN01]).

With this relation given, maximizing the success is equal to increasing the gradient of the success function with respect to changes X in the production. For a constant price p the following relation results with Φ_{cum} the cumulated probability:

3.22

$$\frac{\partial \Gamma}{\partial X} = p \frac{\partial \Phi_{cum}}{\partial X} - \frac{\partial C}{\partial X}$$

It is hence important to determine the probability change of achieving a certain level of performance as well as the associated change in cost. For many changes that can be conducted the effects are difficult to estimate because errors are interrelated. This is particularly true for costs as they can be subject to individual capabilities of a production.

If it comes to the selection of matching parameters, the performance change can be determined for a similar cost level of all the tolerances and the most important tolerances identified. For the laser lens example, image quality on axis is considered and the tolerances in Figure 35 have been sorted in decreasing order of sensitivity.

Figure 35
Tolerance chart for RMS wavefront image quality of the laser focusing lens

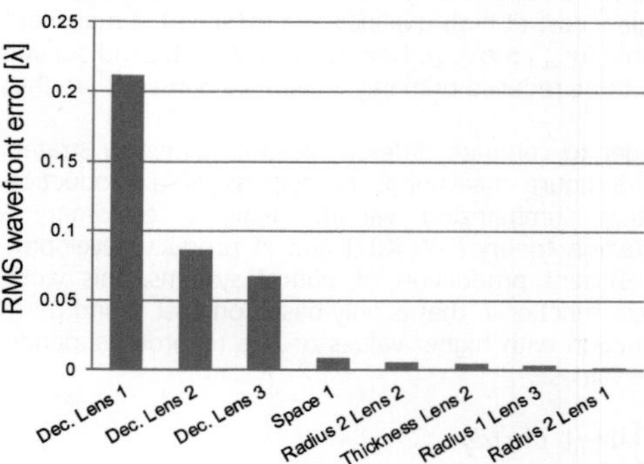

It can be seen that the first three parameters account for a large fraction of the total tolerance. If component one and two are assigned to separate modules a certain amount of compensation is feasible. In addition, especially tight tolerances have high marginal cost and their treatment is most beneficial.

The method is necessarily very approximate and additional effects need to be considered in order to locate a success-optimal definition of modules and tolerances. For this purpose, the next chapter develops a tolerance analysis tool and provides strategies to assign optimal tolerances.

3.4   Summary and conclusion

Combinatorial assembly is a compensation strategy for a small series of systems based on characterization and optimization of component combinations and modules. In this chapter the principles of combinatorial assembly, automatic model-based optimization of component selection to increase a system's performance and a model to assist during the application of combinatorial assembly were developed.

It turns out that the application of combinatorial assembly needs to be planned before production takes place in order to realize matching parameters and employ compensation. Planning the application (finding suitable modules and tolerances) of combinatorial assembly is often not easy due to the number and complexity of performance criteria in optical systems. Reducing the problem to a simple linear model is very helpful highlighting that the complexity lies in the number and interplay of performance specifications. This is illustrated in Figure 36 for a selection of optical systems with increasing complexity for the application of combinatorial assembly.

Figure 36
Overview of typical optical systems and application of combinatorial assembly

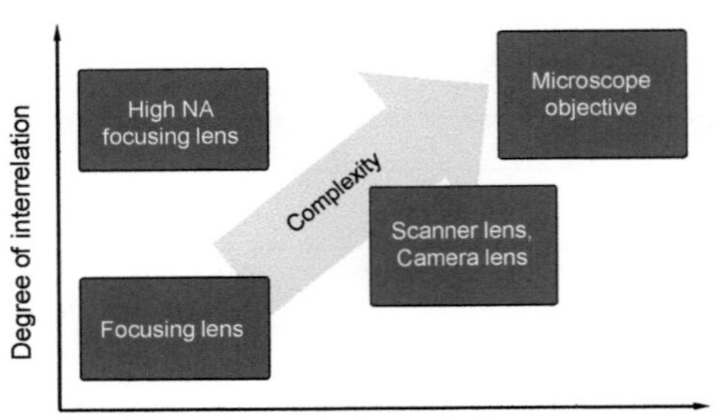

Compensation of single specifications such as spherical aberration in laser focusing lenses is comparably easy, while multiple specifications can pose greater difficulty. If errors can be treated separately without interference multiple specifications can be handled. Examples are focusing systems with high numerical aperture (NA) where centration errors and spherical aberration are critical or astigmatic systems with separable specifications for sagittal and meridional planes. If the performance of optical systems is influenced by virtually every system parameter as in complex imaging systems the application of combinatorial assembly is most difficult.

While simple compensation models are helpful to understand the application of combinatorial assembly, an exact tolerance analysis is indispensable to realistically determine performance increase and applicability of the method for particular optical systems. This will be the subject of the next chapter.

4 Tolerance analysis and assignment

Application of combinatorial assembly requires an analysis tool that can accurately predict the performance of optical systems and allows studying the effects of tolerances. The degree of achievable compensation needs to be determined in order to evaluate the assembly procedure and decide upon its usefulness. Currently available methods are restricted to continuous compensation (e.g. alignment). A tolerance analysis taking discrete compensation of optical systems into account does not exist.

4.1 Tolerance analysis concept

The development of a tolerance analysis method needs to focus on practical usability. It must be applicable to a wide range of problems, integrated into common simulation software and fast enough not to impede the developer more than necessary. The following table summarizes the most important features.

Table 3
Tolerance analysis
requirements

Requirement	Description
Arbitrary parameters/attributes	Any lens parameter, surface type, component position and orientation
Realistic tolerance distributions	e.g. Gaussian, parabolic, user-defined
Variable production runs	Small to medium volume (n=1..50)
Variable number of modules	No principal restrictions, practical limit due to computation (m=1..20)
Additional compensators	e.g. alignment, back focus compensation
Secondary parameter	Include effects from uncharacterized parameters
Arbitrary performance criteria	Any ray-tracing criterion (Basic system parameters and image performance)
Integration in lens design software	Seamless use with sequential commercial ray-tracing environment

The consideration of arbitrary parameters includes the selection of different parameter types (e.g. curvatures, thicknesses and centrations) and the number of parameters and attributes. No restrictions shall be made in this respect and combinatorial assembly of arbitrary modules with corresponding tolerance distributions shall be principally possible. Simplifying assumptions concerning the shape of tolerance distributions can, however, be allowed. Very important is the possibility to account for scaling effects and secondary parameters (tolerances). The larger the production run, the better will be the matching and accounting for these differences needs to be a major feature of the tolerance analysis. A user might want to know the minimum number of assemblies required to render combinatorial assembly a useful strategy. Similarly, secondary parameters which can influence the compensation effects need to be accurately analyzed and virtually every performance measure of ray-tracing environments shall be usable.

4.1.1 Statistical quality prediction

It is the purpose of every tolerance analysis to determine the expected quality of a completely assembled system or a series of systems. Necessarily, the result is either an entire statistical performance distribution or a single value representing the performance of a certain level of confidence. The most important strategies to perform such analyses are discussed in chapter 2.3.1. Not all of them are suitable for the tolerance analysis of combinatorial assembly as it is based on compensation of tolerance effects which are induced by parameter deviations. These effects therefore need to be part of the tolerance analysis and exclude sensitivity analysis. In principle, three possibilities exist to perform the analysis:

1. An analytical description of the tolerance distributions and the system can be used to determine expected mean, variance and higher moments. Regular compensation has been included in such models by some researchers [ADAM87] which could be expanded to account for matching-specific compensation effects.

2. Numerical calculation of combinatorial assembly can be performed using Monte Carlo analysis. This requires the inclusion of the matching process in the tolerance analysis.

3. Differential ray-tracing can be used to substitute Monte Carlo analyses under the restrictions of this particular method.

While analytical analyses are less time consuming, the Monte Carlo approach guarantees the highest possible precision. No simplifications or assumptions need to be made and Monte Carlo types of analyses are very common in the optics community. The only drawback of Monte Carlo is the comparably long computation time. Differential ray-tracing can be used to reduce computation but is only applicable if compensation can be described by the linearization that is assumed. The numerical analysis can build upon optical systems simulation software using the evaluation tools during the analysis process. Despite the large computational effort of the numerical analysis, Monte Carlo analysis is more suitable to the needs than an analytical approach. Due to its versatility and accuracy, this approach will be used and tailored to combinatorial assembly.

The analysis requires the evaluation of an entire production series in a single Monte Carlo run instead of just one random system as combinatorial assembly always acts on a set of systems. Figure 37 illustrates a single analysis run which determines the performance of a series of n systems by generating components with randomly varying parameter values and application of a component mapping: the matching. The number of modules C is denoted m while the $x_{i,j}$ represent the different attributes and q the performance.

Figure 37
Monte Carlo run of
a single matching

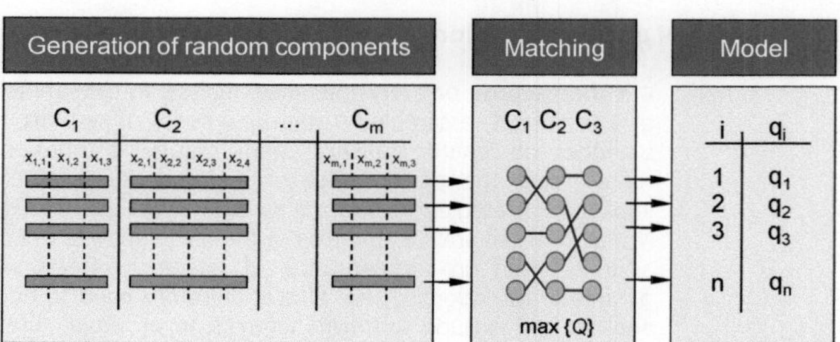

As during standard Monte Carlo analysis, random parameter perturbations are generated following the parameter-specific tolerance distribution law. Instead of using these random numbers in their order of appearance as during Monte Carlo analysis and evaluating the system, modules built of random parameters are matched such that suitable components will be assembled. The same optimization techniques as described in chapter 3 will be employed to find optimal selections of the random components. Due to the randomness of component attributes, matching will yield a different result in every run. That is, matching will generally improve the quality of a system but will not necessarily produce the same level of performance for every system. More importantly, the simulation of the matching will yield multiple systems and one simulation run will be different from every other. It is therefore necessary to build the tolerance analysis around this fact and generate a statistically representative result.

Generating a statistical performance distribution of the series production of matched optical systems requires a two-step approach. First, the optimal matching needs to be performed for a single series production resulting in the performance values q_i of the assembled systems. Then, the simulation of matching a single series needs to be repeated in order to gain a statistically representative performance distribution. The complete procedure is shown in Figure 38.

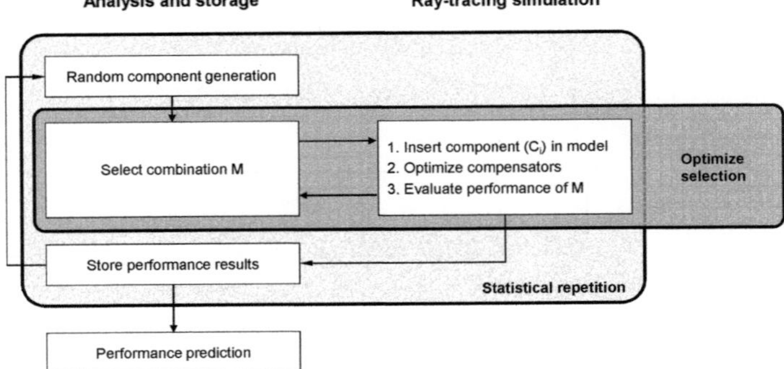

Random components C_i (or alternatively modules A_i) consisting of random parameters x_i are inserted as objects into the ray-tracing model and the optimal configuration is found by iteratively testing different combinations represented by the selection matrix M. This is the same as described in chapter 3 but with random values instead of measurement results. Repetition of this process allows for statistical evaluation.

4.1.2 Tolerance analysis with secondary tolerances

In general, not every parameter will be used during combinatorial assembly. Often, a large number of parameters are secondary tolerances, and only few parameters are characterized and used to select best matches. A tolerance analysis would be far from complete, if only matching parameters are considered and secondary parameters would be neglected. Again, two alternative solutions exist:

- a sequential approach where optimal matching is determined first and secondary tolerances disturb this solution and

- an integrated approach where the optimal matching is influenced by the mean deviation caused by secondary tolerances.

The sequential approach can be realized with an additional tolerance analysis (e.g. Monte Carlo analysis). Based on the optimal matching the effects of remaining parameters are added and in the case of Monte Carlo randomly perturbed in order to estimate the effect of secondary tolerances on the optimal matching solution. Alternatively, the optimal matching selection for a fixed set of components is performed for a number of random secondary tolerances. That matching selection is later chosen, which performs best under all possible situations. Thus, the selection is optimal under the influence of secondary tolerances. The computational effort for the first method is determined by adding a tolerance analysis for every matching selection step, while for the second option matching selection needs to be performed multiple times. Figure 39 shows a comparison of the two strategies.

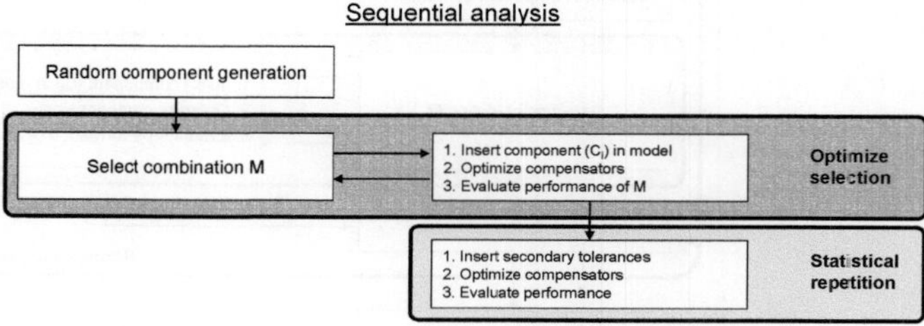

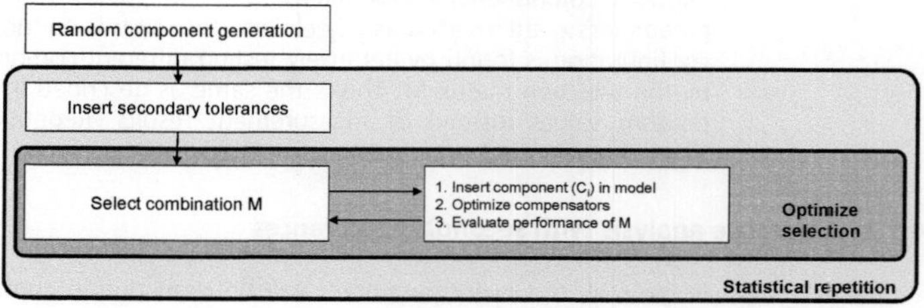

Figure 39
Comparison of sequential and integrated analysis of secondary parameters

If computation time is expressed in terms of the time t_{eval} required to evaluate a system and the time t_{comp} required to find the optimal selection, the total time roughly calculates as

$$t_{seq} = n \cdot s \cdot \left(t_{eval} + t_{comp} \right)$$

$$t_{int} = n \cdot t_{comp} + n \cdot s \cdot t_{eval}$$

As before, n is the number of systems during combinatorial assembly and s is the number of repetitions. Hence, the second variant requires a significantly larger time to calculate. As it is questionable whether or not such an effort would be justified only the first method is implemented.

4.2 Computer-based tolerance simulation

Design and tolerance analysis of optical systems is usually based on ray-tracing calculations performed on a computer. Integration into such a framework is necessary so that the concept of combinatorial assembly can be evaluated similar to conventional strategies. The developed concept uses Zemax® ray-tracing software as a calculation and modeling engine, a MySQL database and a core program for logic operations implemented in Matlab®. The connection and data exchange between the core program and ray-tracing software is realized through a direct data exchange (DDE) protocol enabling programs running under Microsoft® Windows to communicate with each other. The connection between the core program and the database is established with Java (Figure 40).

Figure 40
Overview of the
computer-based
tolerance analysis
program

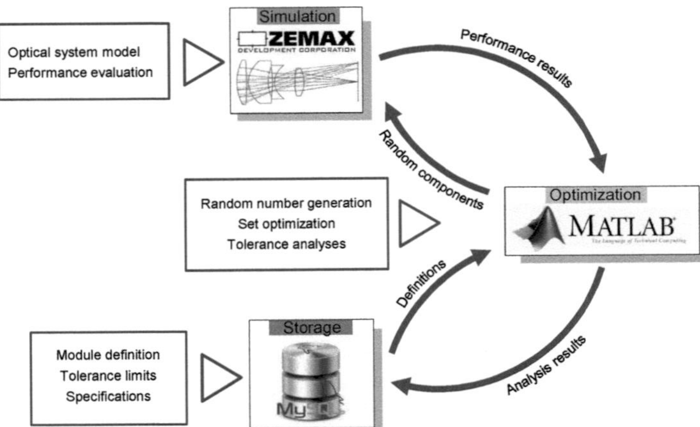

The tolerance analysis is based on a lens design and a merit function entered in the ray-tracing program. Frequently, these can be very similar to the definitions used during design. As the merit function itself is rarely useful on its own, specific operands of the merit function can later be addressed to extract the desired results. Using the ray-tracing software and its merit function allows the user to design and optimize a lens and use the same software for analysis of combinatorial assembly. Every parameter entered in the model can be used during the analysis and any possible number of entries of the merit function can be evaluated. It is thus possible to study the effect of combinatorial assembly on different performance criteria.

Definitions of component types, subassemblies or modules and their according attributes are conveniently entered in the database tables and are accessed by the core program to determine the parameter values that need to be exchanged. Relevant data are the decomposition of the lens design into components, the number of components and tolerance distributions as well as quality criteria used during the optimization process. The use of a database allows a structured definition of the problem in an object-oriented way.

Separate Tables are set up for general data on the analysis (main table), component definition and random parameter values. Figure 41 shows the structure of the database.

Figure 41
Database structure
of the tolerance
analysis program

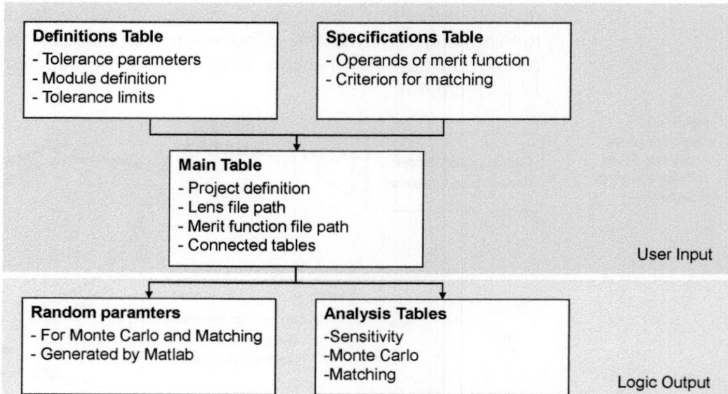

The main table contains project related information such as name, file names of the relevant lens and merit functions as well as the names of related database tables. To start a new project, database tables are automatically created from the core program and data and definitions can be directly entered in the database. These definitions provide information about the tolerances under consideration and are looked up by the logic during analysis, entered in the ray-tracing and results stored. Figure 42 shows an example lens where, modules A and B are defined and a secondary tolerance (SEC) given. The parameters refer to specific surfaces in the lens design model and state thickness (TTHI), radius (TRAD) and surface decenter (TPAR2) tolerances alongside with nominal values.

Figure 42
Definitions table of
the database
(example)

ID	Module	Type	Surface	Adjust_Surf	Value	Tol_min	Tol_max
1	SEC	TPAR2	2	NULL	0	-0.019	0.019
2	A	TPAR2	7	NULL	0	-0.042	0.042
3	A	TTHI	3	NULL	2	-0.05	0.05
4	A	TRAD	4	NULL	11.5	-0.01	0.01
5	B	TPAR2	12	NULL	0	-0.055	0.055
*	NULL	NULL	NULL	NULL	NULL	NULL	NULL

Random components can then be created by the logic according to the definitions and stored in the database. During tolerance analysis, the lens design will be modified by exchanging these random components and testing different combinations. The exchange is performed by substituting attribute's parameter values in the ray-tracing design, resulting performance is evaluated by the merit function and the according values transferred back to the analysis program and then stored. Accounting for adjust surfaces is done by sequentially changing thickness data.

Apart from tolerance analysis of combinatorial assembly, regular Monte Carlo analyses and sensitivity analyses can be performed. Different modes can be selected for the tolerance analysis of combinatorial assembly: rigorous calculation, optimization and heuristic approaches. The program outputs a list of the optimal system configurations after combinatorial assembly together with performance evaluations of these systems. Even if approximate optimization methods are used the final performance is always accurately determined from ray-tracing simulations.

Parameters and modules can also be manually inserted in the ray-tracing program using functions of the core program which accesses the data and placing it in the right place in lens data tables. With this function it is possible to interactively insert modules and observe the results in the ray-tracing environment.

4.3   Tolerancing combinatorial assembly

So far, combinatorial assembly has been introduced as a compensation strategy and a procedure to accurately analyze the predicted performance of optimally matched systems is given. Performance results can be determined for any matching sequence if parameter tolerances, modules and optimization criteria are given. From a lens developer's point of view the analysis tool is very useful to determine the outcome of a particular assembly. What is missing are strategies to assign tolerances to meet performance and cost demands.

Whereas traditional tolerancing concepts are based on change tables and tolerance analyses involving engineering experience, recent approaches attempt automated tolerancing solutions. Similar ideas have been raised for optical systems [YOUN06]. Computer optimization requires that the tolerance analysis model will be evaluated after every step in order to find better solutions. In many optimization algorithms even the derivative to parameter changes are required. Considering the time for regular Monte Carlo analysis, computational effort for tolerance-optimization can easily break all useful bounds and render optimization approaches infeasible for practical applications with dozens or sometimes hundreds of tolerances. From an engineering perspective none of the automated optimization approaches is fully satisfying. They all require a considerable amount of time and are based on data which might not be very accurate sometimes. In addition, insight into the mechanisms and possible solutions is entirely lost and it is therefore questionable whether or not automated solutions to the problem are really necessary and useful.

4.3.1   The tolerance region

In optical design, the performance of an optical system is almost always condensed into a single figure of merit: the merit function. Of practical relevance, however, are individual specifications of system properties (e.g. the resolution at different field points) as these properties can be directly measured and have a clear meaning. Regardless of this evidence tolerancing approaches in the literature focus on combined performance measures [YOUN06] and the multiple specifications are checked after tolerancing. Tolerances are found by manually adjusting tolerances.

Much discussion about the characteristics of the merit function can be observed in the literature and the topography of the merit function is often referred to as the design landscape. While most important during optimization, knowledge about the design landscape is indispensable for tolerancing where only the local characteristics are of relevance. Whereas the merit function is taken as the ultimate performance measure during optimization, it is very uncommon to use the same merit function for tolerancing purposes. Most frequently, tolerancing is performed on optical performance alone, while other constraints are neglected [YOUN06,SMIT85]. As a consequence, some peculiarities of the design optimum need to be taken into account when talking about the tolerance region.

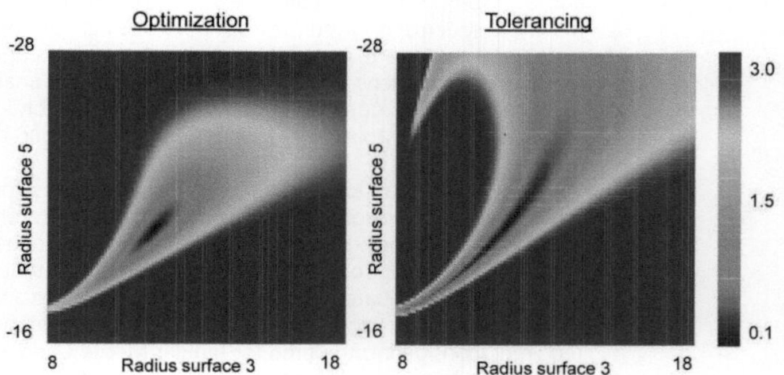

Figure 43 exemplarily shows a two-dimensional merit function. After optimization a local (or global) optimum will be found such that parameter deviations will inevitably degrade system performance, as derivatives are zero in the optimum. However, this is only exactly true for the merit function used during optimization and the parameters that were allowed to vary. Boundary conditions, Laplace operators and parameters that remain fixed during optimization will cause the tolerance region to depart from optimality in some parameters. In addition, the merit function contains entries that might be irrelevant during tolerancing and were only necessary to find a satisfactory design during optimization. As a result the tolerance region of the criteria used for tolerancing is not an optimum, at least not for every parameter. Others report about design centering (a shift in the position of the optimum) indicating similar observations [SZAP10]. This fact needs to be borne in mind as it implies that a small parameter deviation can actually increase the performance and thus allow compensation. Combinatorial assembly can make use of the long and stretched valleys and use the flexibility to tolerance changes of the design. However, the particular shape of the merit function due to its root sum square nature, rendering the merit function necessarily nonlinear, will often conceal compensation potential during sensitivity analysis as pointed out before.

4.3.2 Matching tolerances

As a basic tolerancing rule for combinatorial assembly it seems logical to choose tolerance ranges such that tolerance regions are overlapping as mentioned in section 3.3. The idea is illustrated in Figure 44.

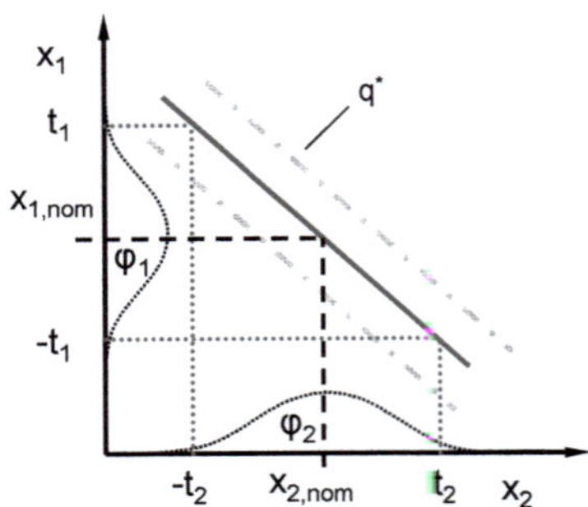

Figure 44
Perfect match of tolerances for two parameters

Perfect matching can be obtained in the limit of an infinite number n of available components if corresponding tolerance probabilities always match. This can be expressed as the following integral, where f represents the function describing the optimal curve of perfect compensation (e.g. $q^*=0$) and T the tolerance interval.

4.2
$$\int_{T_1} \varphi_1 f_1(x_2)\,dx_1 = \int_{T_2} \varphi_2\,dx_2$$

For any symmetrical tolerance distribution φ and linear performance criteria this is fulfilled if the tolerance limits t_i multiplied by their sensitivity S_i equal each other.

$$4.3 \qquad S_i t_i = const \quad \forall i$$

If the tolerance regions of parameters do not overlap with the performance function, perfect compensation is no longer feasible and the ideal matching result will change. For the uniform and Gaussian case of a linear performance function $q*$ the curves are given in Figure 45 (right). It is assumed that the optimal matching is attained by matching sorted components and that the mean of the distribution coincides with the optimal matching curve.

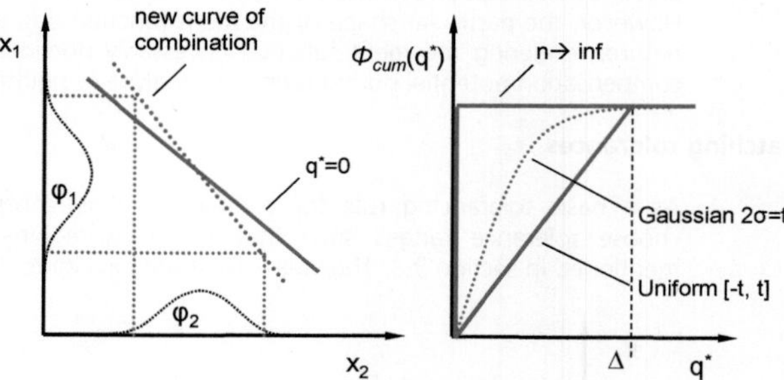

Figure 45
Imperfect matching
of tolerances

For an incomplete Matching the quality limit (e.g. Φ_{cum}=99%) will be shifted to lower quality levels. The resulting Δ for the uniform distribution directly corresponds to the distance at the tolerance limits.

$$4.4 \qquad \Delta = \left\| S_1 \right| \cdot t_1 - \left| S_2 \right| \cdot t_2 \right\|$$

The mismatch then impacts the performance and, in general, the resulting cumulated probability function Φ_{cum} of quality $q*$ can be calculated as the integral of the probability function along the new matching curve (dotted line). In the case of a linear performance relation the matching line can be assumed to follow a line connecting similar levels of the probability functions.

$$4.5 \qquad \Phi_{cum}\left(q*\right) = 2 \int_0^{x_1\left(q*\right)} \varphi_1 \, dx_1$$

The upper limit for integration can be derived as

$$4.6 \qquad q*\left(x_1\right) = \frac{\Delta}{t_1} \cdot x_1 \Leftrightarrow x_1\left(q*\right) = q*\left(x_1\right) \cdot \frac{t_1}{x_1}$$

The resulting Φ_{cum} for a uniform distribution is a linear function from zero to Δ while for a Gaussian distribution with standard deviation σ, an error function (*erf*) results. As said before, the distributions are centered to each other.

$$\Phi_{cum}\left(q^*\right)=\frac{2}{\sigma\sqrt{2\pi}}\int\limits_{0}^{x_1\left(q^*\right)}e^{-\frac{x_1^2}{2\sigma}}dx=\frac{1}{\sqrt{\sigma}}erf\left(\frac{x_1\left(q^*\right)}{\sqrt{2\sigma}}\right)$$

4.7

Shifted tolerance curves with the means of the distributions decentered towards the ideal matching will increase the probability of an error unnecessarily and are therefore never useful. Depending on the tolerances and their cost impacts it can however be useful to allow imperfect matching. In this case one matching parameter tolerance is widened to reduce its relative cost of tolerances while the tolerance of another tolerance is increased.

Perfect compensation as described above is indeed possible if matching is executed with a large number of available components and if the probability functions match. In the linear case matching is only perfect for an infinite number of components and consequently matching will in reality never be perfect even in the linear case.

Several effects cause matching to depart from the ideal situation and necessitate thinking about the optimal tolerance allocation. For tolerancing purposes it is important to look at the changes of the performance function if tolerance limits are varied. That is the partial derivative of a performance level with respect to the tolerance. Different cases can be distinguished which will be illustrated for linear matching.

-   performance changes for perfect matching due to small, but finite number of systems n
-   performance changes due to a mismatch of the tolerances
-   performance changes for mismatched tolerance limits

First, matching tolerance regions will be considered, and discussed how the tolerance limit changes the expected performance in the linear case for uniform and Gaussian tolerance distributions. The difference of ideal matching (infinite n) to the actual performance limit can be described as the average distance to the optimal curve (the arithmetic optimization measure Q^*_1). When tolerance limits are changed, the average distance to the optimal curve will change as the scattered points of module combinations are further apart. As a consequence the performance will change too. For the uniform and Gaussian case the change of a performance criterion will be (approximately) linear with the tolerance limit.

Figure 46 shows the performance q^* for simultaneously changed tolerance limits of two matching parameters of the laser lens example. The parameters are decentrations of lens 2 (t_1) and lens 3 (t_2) and their sensitivity to Zernike tilt is considered. For $t_1=t_2=25$ µm the changes calculate as $S_1t_1=-0.218$ and $S_2t_2=-0.167$. If tolerance one is changed to meet the matching criterion, $t_1=19$ µm satisfies the condition that $|S_1|t_1 = |S_2|t_2$.

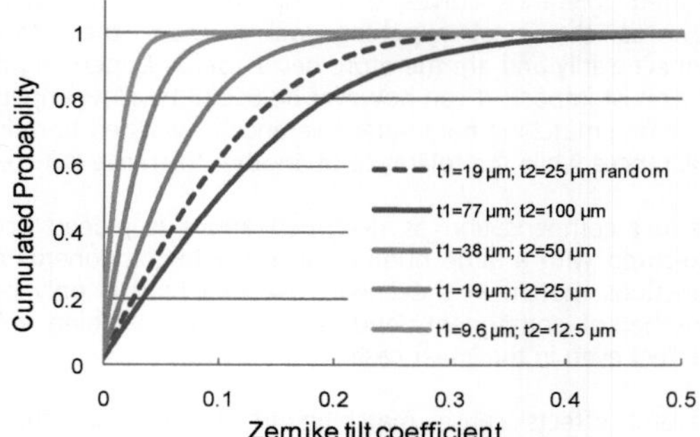

Figure 46
Tolerance analysis for matching tolerance limits and Gaussian tolerance distribution (n=20)

The graph illustrates the cumulated probability of achieving a performance level if the tolerance limits are changed in proportion. The differential with respect to tolerance limits is constant and can be regarded an effective sensitivity that is referenced to the perfect matching. This effective sensitivity can be determined by running a tolerance analysis of combinatorial assembly with matching tolerance limits instead of a sensitivity analysis.

For non-linear relations the picture will change and the sensitivity will no longer be constant for different tolerance limits as matching can never be perfect and the departure from full compensation will also be a function of the tolerance width. Due to higher order errors, the quality drop with loose tolerances will be quicker.

If tolerance distributions do not match, compensation is reduced even for an infinite number of components as has been illustrated above. For a small production series uncertainty will add to the theoretically achievable limit. This is illustrated in Figure 47 where two parameters can compensate each other but only one tolerance is changed.

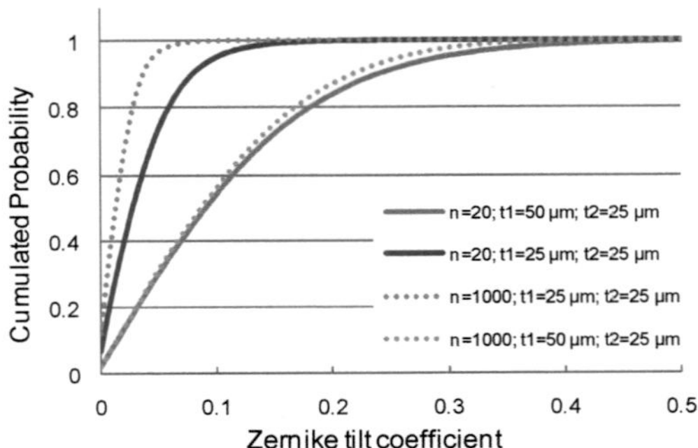

Figure 47
Tolerance analysis of
mismatched
tolerances

One could expect that the difference to the theoretical curve is similar to the difference due to uncertainty in the case of matching tolerances. However, this is not quite true and the uncertainty effect is in fact reduced such that for imperfect matching performance is closer to the theoretical curve. Note for example how there is almost no difference between n=1000 and n=20 for the least match of tolerances (t_1=50 µm, t_2=25 µm). Hence it is not possible to reduce a change of only one tolerance to a simple sensitivity.

If the result of imperfect matching is compared to random assembly and perfect matching the following graph can be derived (Figure 48).

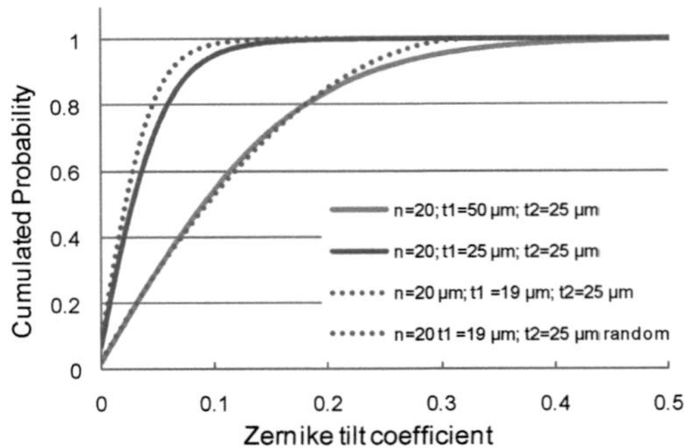

Figure 48
Comparison of
mismatched
tolerances with
random assembly
and perfect
matching of
combinatorial
assembly

The curves for random assembly with 19 µm and 25 µm tolerances are very similar to the imperfect matching for 50 µm and 25 µm tolerances. Hence, even for an extreme deviation from perfectly matching tolerance limits, a decent performance increase is obtained.

If imperfectly matched tolerances are scaled in proportion the same law as stated above for matching tolerances applies. As illustrated in Figure 49, reduced tolerance limits decrease the quality deviation in exact proportion to the change of the tolerances if their ratio remains constant.

Figure 49
Scaling of
imperfectly matched
tolerance regions

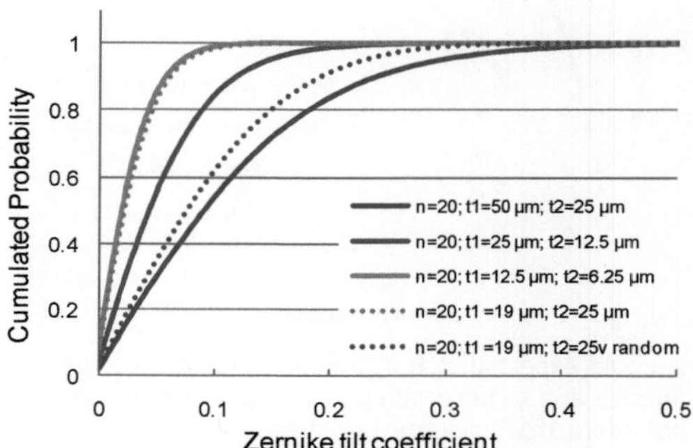

Note how perfect matching of tolerance 25 µm is almost exactly equal to an imperfect match of 12.5 µm and 6.25 µm. It can thus safely be concluded that only small departures from perfect matching should be allowed.

4.3.3 Scaling with production volume

An important feature of the tolerance analysis is calculating the degree of compensation as a function of the production volume n. As the number of possible combinations increases, the possibility to approach perfect compensation will grow and the quality distribution of a set of systems will narrow. If a performance limit is specified it is of particular interest to determine the size of a production run that will yield the desired precision. While the required number of components can be determined by experimentally changing the values in simulation, knowledge of the underlying scaling laws will be very insightful.

Figure 50 shows the difference in theoretically achievable compensation for a non-linear performance measure (merit function) and symmetric tolerance distributions for different production volumes.

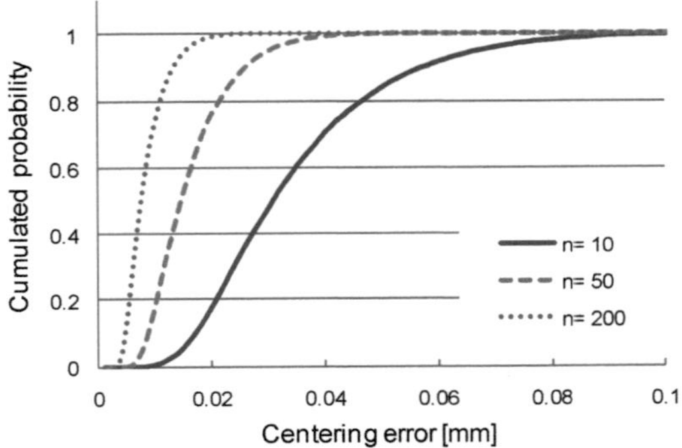

The curve of optimal combination as shown in Figure 44 can be regarded a best fit to possible component combinations during combinatorial assembly. As with a changed tolerance limit, the proximity of the solution to the ideal curve depends on the number of available module combinations. For the mentioned case the optimal matching configuration is given by the sorted components and the deviation of optimal curve is then given by

$$\Delta q\left(n_j\right)= \Delta q\left(n_i\right)\cdot \sqrt{\frac{n_j}{n_i}}$$

4.8

For every performance value of a production run of size n_i the corresponding point of the quality curve for a production volume of n_j is found by multiplying a constant factor. The result can be explained based on the average mean distance between samples which decreases in inverse proportion to the number of systems n. It is furthermore independent of the number of dimensions m as long as a problem with a perfect compensation solution is considered (e.g. a line in 2D or 3D). If multiple parameters can be used for a single compensation the effective density of samples increases and so does the performance. The number of samples in one direction will then grow exponentially. Scaling, however, stays the same as the ratio of the exponential increase will remain constant for every particular case.

For more complex problems scaling cannot easily be determined analytically. However, if a similar scaling law as the one in equation 4.8 is assumed a generalized scaling law can be fit to the data in order to make predictions.

4.3.4   Cost-optimized tolerances

Whereas different performance measures can be combined into a single figure of merit to evaluate a system's overall performance, costs cannot generally be incorporated into this figure of merit. This is only possible if a true cost-performance trade-off is acceptable. As a consequence it is desirable to find a tolerance assignment that minimizes the costs while staying within prescribed performance limits.

The effects described in the previous section indicate that not only for random assembly but also for combinatorial assembly looser tolerances reduce performance. As we have seen this is true for any production with a (small) real number of systems and particularly true for non-linear performance functions. Hence, it is not generally possible to assign matching tolerances as loose as possible to reduce cost. In addition, tolerances of matching parameters should be given such that tolerance distributions match. While this may work well for more or less symmetrical problems with similar sensitivities, it can render combinatorial assembly inapplicable in extreme cases.

As detailed before, costs can be assigned to assembly processes and to components. Hence, total costs are a function of the component tolerance limits t_i as well as the assembly processes AP_i. Because the cost of assembly processes have been fixed before tolerance assignment and component base costs are equal for every assembly sequence only component tolerances need to be considered in a cost-minimization process. As tolerancing is the last step during every iteration of the development procedure the costs of tolerances should always be minimized.

Necessarily, the cost-optimal tolerances are such that performance is barely met as tolerances should be as loose as possible. Hence, cost-optimal tolerances produce the minimum allowed performance. If curves of constant performance and constant cost are plotted as a function of tolerances, the cost curve for combinatorial assembly will be further away from the origin than the curve for random assembly. As a consequence the tolerances to achieve the same level of quality can be larger and the costs are reduced as mentioned before. Figure 51 shows curves of constant cost C with $C_2 > C_1$ and curves of constant performance q ($q_{CA}=q_{RA}$) for random assembly and combinatorial assembly. The tighter the tolerances are, the higher the performance but the higher also the cost.

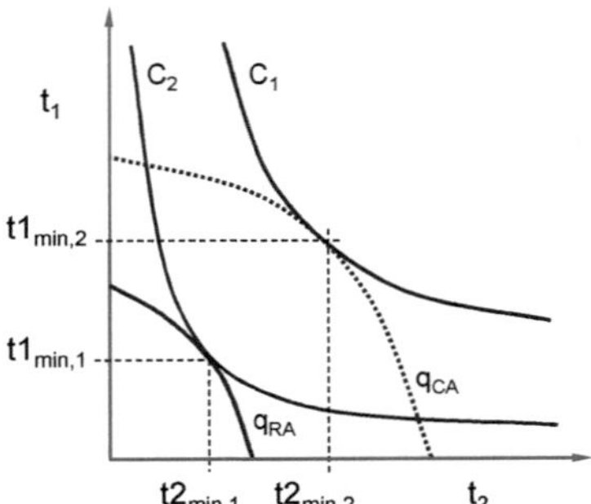

The minimum costs are obtained at the point where cost and performance curve touch each other. There, the partial derivatives of cost and performance with respect to tolerances are equal. Let C denote costs and q_c a curve of constant performance. Then:

4.9

$$\frac{\partial C}{\partial t_i} = \frac{\partial q_c}{\partial t_i}$$

The cost-optimal tolerancing as described by Adams and originally introduced by Grammatin and Kundeleva uses Lagrangian Multipliers to derive cost-optimal tolerances for worst case estimates of linear performance effects [ADAM87]. This concept shall here be expanded for combinatorial assembly taking the effects described in the previous section into account. In particular, the effect of simultaneously scaled tolerances with matching limits will be important. As demonstrated, the quality after combinatorial assembly depends linearly on the scale of the tolerances no matter if matching is perfect or not. In addition, the degree of mismatch between tolerances has an influence. This fact can however be neglected for the moment assuming that tolerances will always match. The idea is to introduce an effective sensitivity for matching parameters such that the standard performance estimates can be used:

4.10

$$q_{WC} = \sum_i |S_i| t_i$$

$$q_{RSS} = \sqrt{\sum_i S_i^2 t_i^2}$$

With the determining function F and the Lagrangian Multiplier λ

4.11

$$F(\underline{t}) = C(\underline{t}) + \lambda \cdot q(\underline{t})$$

The method makes use of the fact that all the partial derivatives of $F(t)$ should be zero. With an inverse cost-law

4.12

$$C(t) = \frac{1}{t}$$

this delivers cost-optimal tolerances for the worst case and RSS estimate if they are

4.13

$$t_{i,WC} = \frac{q_{WC}}{\sqrt{S_i}\sum_j \sqrt{S_j}} \quad , \quad t_{i,RSS} = \frac{q_{RSS}}{\sqrt{S_i^2 + \sum_{j\neq i} S_j^2 \left(\frac{S_i}{S_j}\right)^{4/3}}}$$

In this equation the S_i and S_j are now the sensitivities of secondary tolerances as well as the effective sensitivities of matching tolerances that will be derived. Note that the summation for the worst case goes over all the sensitivities. Incorporating matching effects requires to find the effective sensitivities and to guarantee a fixed ratio between tolerances to realize matching. The following should then hold true for the worst case scenario in order to retain matching tolerance widths.

4.14

$$|S_1|t_1 = |S_2|t_2 \Rightarrow |S_1|\frac{1}{\sqrt{\lambda|S_{eff,1}|}} = |S_2|\frac{1}{\sqrt{\lambda|S_{eff,2}|}} \Rightarrow S_{eff,1} = \left(\frac{S_1}{S_2}\right)^2 S_{eff,2}$$

Here, S_{eff} denotes the effective sensitivity while the S_i denote the original sensitivities. For the RSS estimate the relation between effective sensitivities turn out to be

4.15

$$|S_1|t_1 = |S_2|t_2 \Rightarrow S_{eff,1} = \left(\frac{S_1}{S_2}\right)^{3/2} S_{eff,2}$$

Bringing the effective sensitivities into the right proportion is important to obtain cost-optimal tolerances. Given the above relations, the effective sensitivity can be determined by implementing the effective sensitivities in the performance equations as given in 4.10, yielding

4.16

$$S_{eff,WC,1} = \frac{q_0}{\left(\left(\frac{S_2}{S_1}\right)^2 t_2 + t_1\right)} \quad , \quad S_{eff,RSS,1} = \frac{q_0}{\sqrt{\left(\frac{S_2}{S_1}\right)^3 t_2^2 + t_1^2}}$$

The effective sensitivity for the second tolerance can be calculated analogously and the change in quality q_0 can be observed from a tolerance analysis of combinatorial assembly with tolerances t_1 and t_2. The original sensitivities S_i are obtained from regular sensitivity analyses. Unlike regular tolerances the effective sensitivity will therefore depend on the number of assembled systems n as the observed quality changes too. In addition, the same formulation can be used for imperfectly matched tolerances as the scaling law is the same. Again, only the effective sensitivities will change.

For the given example and a 98% limit $q_0 = 0.1$ can be determined from Figure 46 for 19 µm and 25 µm tolerances. The effective sensitivities then calculate as $S_{eff,1}=0.00297$ and $S_{eff,2}=0.00174$ for the worst case and limit and $S_{eff,1}=0.00356$ and $S_{eff,2}=0.00239$ for the RSS estimate. With these effective sensitivities the Lagrangian Multiplier method can be applied in its known form such that the tolerances can be determined with equation 4.13.

If the laser focusing example is expanded by a third tolerance with $S_3=0.0148$ (centering of lens 1), a limit of $q=0.5$ demanded for the Zernike tilt coefficient, and costs inverse proportional to tolerances, the following cost-optimal tolerances can be derived: $t_1=42$ µm $t_2=55$ µm, $t_3=19$ µm as opposed to $t_1=18$ µm, $t_2=21$ µm, $t_3=14$ µm without matching. Costs drop from 0.175 to 0.095 by a factor of 1.84. For the RSS case tolerances arrive at in $t_1=67$ µm $t_2=88$ µm, $t_3=26$ µm as opposed to $t_1=32$ µm $t_2=38$ µm, $t_3=22$ µm.

With a comparably small effective sensitivity of the matching parameters, the result has some interesting implications. Tolerances of matching parameters should ideally be large, while secondary parameters should have tight tolerances to achieve cost-optimality. This in turn can be beneficial for the matching itself, as higher order effects disturbing compensation can be reduced by tighter secondary tolerances. It also highlights again that matching should be performed on costly tolerances. (The effect was exaggerated in this example because t_3 wouldn't technically be a secondary parameter as it has a high sensitivity). If applied properly, matching can help to equalize tolerances. Figure 52 shows the result of the tolerance analysis.

Figure 52
Simulated performance for combinatorial assembly with and without secondary tolerances

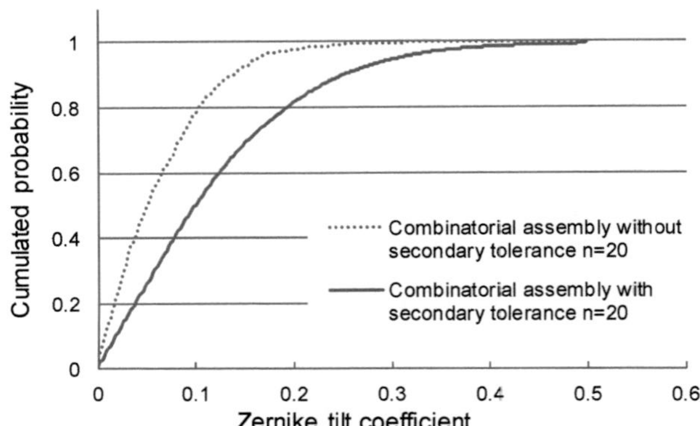

Evaluation of the so obtained initial trial with exact tolerance analysis determines whether or not performance requirements can be met with the selected strategy. If further cost reduction appears feasible, tolerances can be iteratively changed by successively updating the effective sensitivities for the newly determined tolerances. The results stated above are valid for worst case and RSS estimate for costs that vary in inverse proportion to tolerances. The result will necessarily change if other cost models are applied. The principles of calculation however remain the same.

4.4   Summary and conclusion

Analyzing the effect of combinatorial assembly on system performance is very important to judge whether or not it is useful to apply such an assembly strategy. The derived tolerance analysis concept builds on ray-tracing simulations and Monte Carlo analyses as these are the most exact and versatile performance analyses for classical optical systems. Combinatorial assembly is incorporated by selecting optimal module combinations during Monte Carlo analysis. In order to gain a statistically significant result, Monte Carlo analyses of a single combinatorial assembly are repeated and if the number of runs is large enough the results can be very accurate.

In reality, parameter values are determined through measurements (or, if neglected, not measured at all) and consequently possess an uncertainty. This in turn will influence the results and needs to be taken into account as a secondary parameter. The developed concepts to account for these effects are integrated into a tolerance analysis software making use of ray-tracing, logic operations and data storage. Virtually any optical system that can be modeled and evaluated with the ray-tracing software can be analyzed.

To assist the assignment of tolerances the effects of tolerances on cost and performance are illustrated. The most important conclusions are that tolerances should almost always match and that their tolerance limits impact performance for small series production such that a cost-optimal set of tolerances can be found with effective sensitivities.

Mismatch of tolerance limits is sometimes unavoidable if parameter sensitivities are very different or if manufacturing are reached. In addition, the cost advantage may be lost. This can reduce the application of combinatorial assembly in practice as can the influence of secondary parameters on system-performance. It is also possible that secondary parameters turn out to be critical and should be used as matching parameters too.

Making changes to the optical design may help to reduce these undesired effects and the next chapter deals with strategies to reduce the effects of secondary tolerances and increase the match between tolerances.

5 Design for combinatorial assembly

Interrelations between optical performance specifications and the effects of secondary tolerances have proven to be one of the main difficulties in the application of combinatorial assembly. In addition, tolerances should match such that good compensation can be achieved. It is a logical consequence to think about design strategies that yield altered lens designs and fully use the potential of combinatorial assembly by finding suitable locations in the lens design landscape.

Designing optical systems is almost exclusively performed using automatic correction and the optical merit function has developed its status as the ultimate measure of design performance. Of the interpretations in the literature, two are of notable interest: Bociort describes the global merit function landscape as a network of local minima and maxima [VTUR09] while Isshiki pictures local optima along a chain in multidimensional space [ISSH07]. The structure of the design landscape gives rise to a few interesting insights (Figure 53).

Figure 53
Different solutions in the design landscape and their application

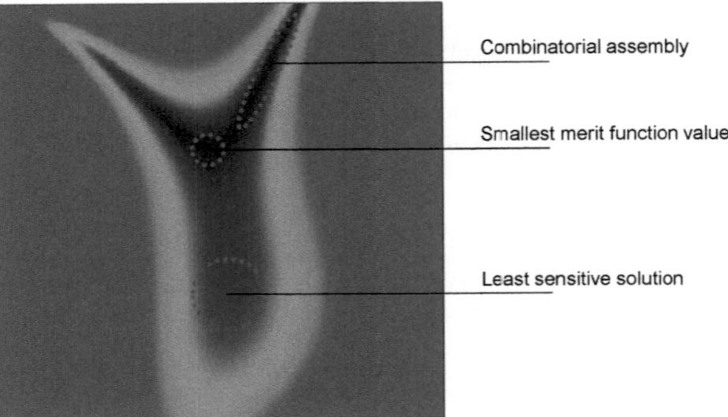

While some local minima are deeper than others, the vicinity around minima is also different. As tolerances and environmental influences may cause the design to move away from the true minimum this can be an important factor. A structure as described by Isshiki and Gross with long and stretched valleys can be well suited for combinatorial assembly. Then, there is a certain degree of freedom in which the design can move without sacrificing performance too much. This has been used throughout the examples that were given but without looking at the design landscape. A lens design may benefit differently from combinatorial assembly, depending on its location in the design landscape.

5.1 The effect of combinatorial assembly on desensitization

If a system is said to be insensitive it is expected that the performance of the system does not change significantly when parameters are disturbed (e.g. due to tolerances). Desensitization is thus the finding or generation of design forms that exhibit only small performance degradation under the presence of parameter perturbations and deliver the best as-built performance. Typically, desensitization focuses on imaging properties as these are the most critical specifications in an optical system. Most of the research has been performed on reducing tolerance effects of decentrations, tilts and other mechanically induced errors to facilitate assembly and reduce manufacturing costs. Desensitization can principally be achieved by changing

a) only available design parameters

b) an increased number of design parameters by adding new design elements (additional lenses, aspheric surfaces)

c) a reduced number of *relevant* design parameters by applying different assembly strategies such as compensation

Suppose a design optimum has been found (e.g. minimum of the merit function) what can be expected from desensitization? Desensitization will find a point in the merit function that has a higher value than the former solution, but with lower tolerance effects such that the as-built performance is better than before. Only with great luck it can be expected to find a solution that is both better and less sensitive. If nominal performance and tolerance effects are investigated separately as a function of lens designs, their respective minima will not necessarily coincide. The task is therefore to find the best compromise between nominal performance and tolerance effects. This compromise between sensitivity and nominal merit function can be observed in other works on classical assembly of optical systems as well [ISSH07].

Figure 54 illustrates the balancing of nominal system performance and tolerance effects for a three element laser focusing lens (Figure 14). Designs were varied according to the desensitization method of Isshiki with curvatures and thicknesses variable and different average angles of incidence (root sum square of ray incidence angles on lens surfaces, compare [ISSH07]) are plotted. Tolerance contributions are 98 percentile values obtained from Monte Carlo tolerance analyses.

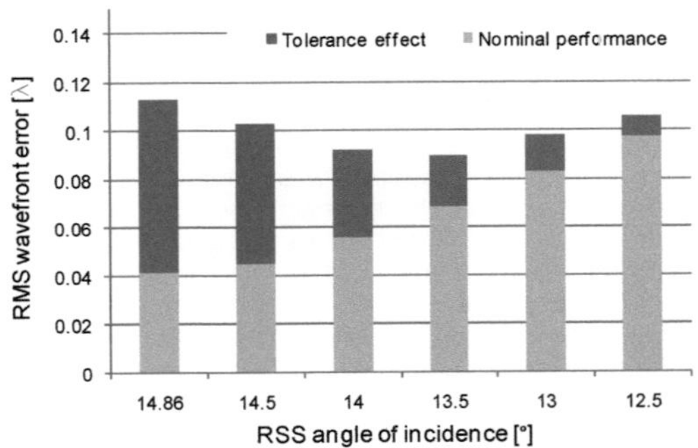

Figure 54
Nominal performance and tolerance induced changes of a three element laser focusing lens

The graph shows that the optimal as-built solution is neither the one with the lowest tolerance effects nor the one with the best nominal performance. A tradeoff between nominal performance and tolerance sensitivity thus needs to be found.

If the effects of separate parameters on performance are observed for different design optima during desensitization it becomes obvious that sensitivities are shifted. Different solutions show very different sensitivity distributions and it is frequently possible to reduce one or more parameters in their sensitivity to a tolerable value, while at the same time sensitivities of other parameters will increase. This is illustrated in Figure 55 for three lens designs of the laser focusing lens (compare Figure 67 and Appendix 1.3).

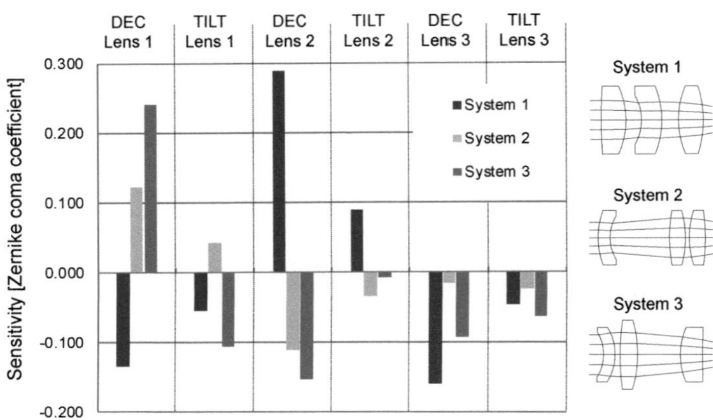

Figure 55
Sensitivities of three design choices against 0.05° tilt and 25 µm decenter of lens elements

In addition to shifted sensitivities it is often possible to specifically reduce the sensitivity of selected surface, for example by reducing the angle of incidence or by making the surface aplanatic.

A simultaneous reduction of all sensitivities seems almost impossible. Instead, desensitization will result in sensitivities which are more evenly distributed over the system thereby reducing the overall tolerance effect. If a root sum square estimate is assumed to represent the estimated performance change this becomes clear. The largest values dominate the performance change and the RSS is minimal for equal changes. If desensitization results in a design with manageable tolerances, this is fine. If however, sensitivities are distributed over the entire system such that every component is somewhat critical, this can be a highly undesirable result. It would then often be easier to concentrate on a few critical tolerances during manufacture or assembly and neglect the remaining tolerances.

This idea has led to the question whether combinatorial assembly could be used in conjunction with desensitization. If combinatorial assembly is applied to very sensitive parameters (e.g. worst offenders) their tolerance effect can be significantly reduced as has been shown in chapter 4. But this is not everything there is to combinatorial assembly. The application of combinatorial assembly – or compensation in general – allows the designer to focus desensitization of a design on parameters which will not be used for compensation. If in doing so, a few very sensitive parameters can indeed be excluded from desensitization it should be expected that the remaining sensitivities can be reduced even further. A different design optimum can then be found for the reduced set of disturbed parameters and hopefully the trade-off between nominal value and tolerance effect can be improved. In reducing only the sensitivity of secondary tolerances, the sensitivity of parameters used during combinatorial assembly may rise which is – at least theoretically – not as critical.

Figure 56 illustrates the idea for three systems from Figure 54 (three element laser focusing lens). While the best compromise of nominal performance and tolerance effect during desensitization of the entire lens requires a significant increase of the nominal value to reduce the tolerance effect, a smaller increase is sufficient to find the optimal compromise for the reduced set of parameters. For this example it is assumed that perfect compensation of matching tolerances is possible and executed for a large number of modules. The combinatorially assembled system can thus be designed to have better nominal performance, whereas a large part of the tolerances is reduced through combinatorial assembly.

Figure 56
Tolerance reduction through combinatorial assembly. Optimized design (OPT), desensitized design (DES) and combinatorial assembly matching (MAT)

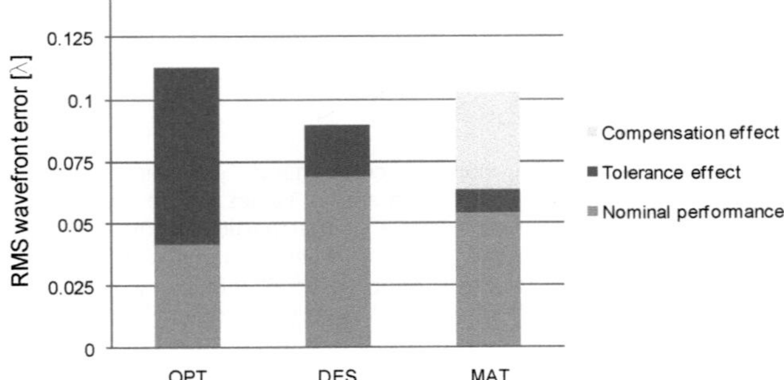

It is important to note that the comparison between purely desensitized designs and designs desensitized for combinatorial assembly depends on the solutions found. It might always be possible to find a desensitized design with even better as-built performance and the combinatorial assembly design would have to be compared against this solution. Similarly a better solution for combinatorial assembly might be found and ideally, global cost-optimal solutions would be compared, but this is practically impossible to realize.

A few more things are necessary for the idea of combining combinatorial assembly and desensitization to work in practice. First, sensitivities must indeed tend to even out during desensitization and can be redistributed such that parameters for selective assembly gain sensitivity, whereas other tolerances loose sensitivity. This can depend on the design form and the constraints imposed on the design. Second, a feasible matching between components must be possible and reduce most of the tolerance induced changes. Ideally, the sensitivities of selective assembly parameters entirely disappear due to matching and secondary tolerances have small sensitivities. As illustrated in chapter 4, this necessitates perfectly matching tolerance distributions and as many available modules n as possible. And third, secondary parameters must not significantly disturb selective assembly. That is, the matching is insensitive to other tolerances and cross-relations are small. For linear performance measures, this is naturally given.

5.2 Design for insensitivity and combinatorial assembly

Two main steps dominate the design of tolerance optimized lens designs for combinatorial assembly:

- searching and amplifying compensators
- desensitization of secondary parameters

With the help of the compensation model introduced in section 3.3 the strategies can be interpreted. The deviation T of a performance specification is given as:

5.1

$$T = \sum_i \underline{c}_i \underline{S}_i^{(1)} + \underline{c}_i \underline{\underline{S}}_i^{(2)} \underline{c}_i^T$$

Again, individual components C_i are represented by a vector $\underline{c}_i$ containing the changes Δx of parameter values x from their nominal values such that only the parameters values of that particular component are allowed to be non-zero. The vector of sensitivities is denoted $S^{(1)}$ and the matrix of second order sensitivities is denoted $S^{(2)}$. If multiple specifications need to be met, a set of equations can be derived that should all equal zero.

The two strategies aim at increasing compensation by

- minimizing higher order effects

- avoiding interrelations with other specifications

- achieving matching tolerances

- placing matching parameter on different modules

- reducing the sensitivity of secondary parameters

Higher order effects reduce the possible compensation while interrelations of specifications complicate the finding of proper tolerances. Matching tolerance ranges is necessary, because the sensitivities of matching parameters can be so vastly different that it is impossible to achieve perfect matching. Tolerances would have to be changed beyond their feasible limits and in these cases additional matching parameters are necessary to mitigate the effect. In addition, compensation through combinatorial assembly is only feasible if matching parameters are located on different components or modules.

5.2.1 Changing sensitivities to match tolerance regions

Whereas desensitization helps reducing secondary tolerance effects that superimpose the positive effect of compensation, adaptation of tolerance regions has the goal to match tolerances in order to fully exploit the compensation potential. With the help of design changes it should be possible to influence the sensitivities of different modules to generate a closer match thereby increasing the potential for matching and reducing the cost of tolerances. Once matching parameters are selected it would be most beneficial to adapt their sensitivities such that their respective effects are at a similar tolerance level (e.g. base tolerance) are equal.

5.2

$$S_i t_i = const.$$

Instead of adjusting the tolerances as described in section 4.3, the sensitivity is changed to create the match. Ideally, sensitivities would be chosen such that a perfect match is created for cost-optimal tolerances.

In order to adapt a design the restrictions arising from matching tolerance regions need to be incorporated into the optical merit function. The possibilities are:

a) a multi-configuration approach

b) restrictions on sensitivities of the merit function

c) restrictions on aberration coefficients or Zernike polynomials

For the multi-configuration approach, configurations with fixed tolerances are created which should ideally match and optimization is then carried out for all configurations simultaneously (Figure 57).

Figure 57
Multi-configuration
matching approach

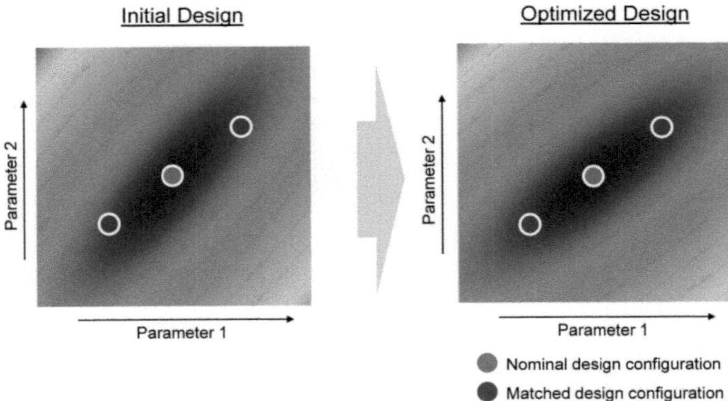

The method has the advantage, that creating configurations which should later match is relatively straight forward. Consider the laser lens example and matching of centrations as discussed during cost-optimal tolerancing. Attempting to match the tolerances of centrations of lenses 1 and 3, three configurations can be defined; one for the centered design and two designs with corresponding decentrations of the lenses. Optimization of radii and thicknesses keeping the back focal length within prescribed limits gives the following result (Figure 58):

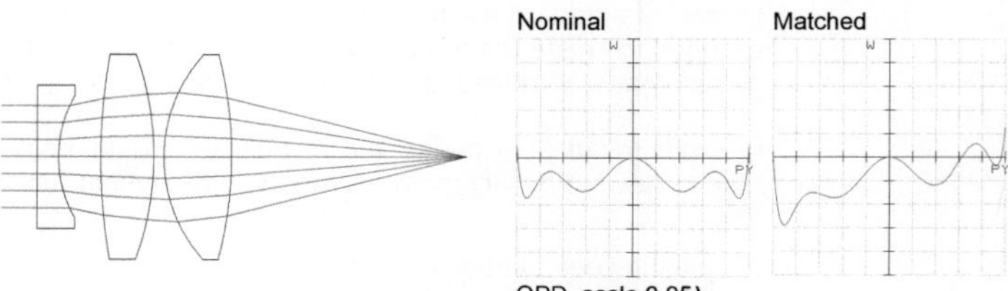

Figure 58
Laser focusing lens
optimized for
matching tolerances
with multi-
configuration
approach

Weights need to be defined for the different configurations, in order to balance performance between nominal and matched configurations. The results in Figure 58 show that the wavefront is not perfectly compensated at maximum tolerance limits but wavefront errors are very small. The optimization process does not converge very quickly due to the large amount of rays traced through the system.

It is also advisable to define not more than two symmetrical configurations per tolerance as otherwise nonlinear effects disturb the optimization. If instead of multiple configurations, sensitivities are directly restricted and weights set in the merit function, sensitivities need to be determined during every optimization step. This approach is also used during desensitization where sensitivty targets are small. Many ray-tracing programs offer this possibility. The two methods are practically identical.

Alternatively, Zernike polynomials can be employed to measure the sensitivity to specific errors. This has the advantage that the designer can select the most important errors and thus simplify the problem as Zernike coefficients are quickly calculated. To incorporate Zernike coefficients, separate configurations for single parameter changes are introduced in the model and the Zernike coefficients evaluated. This is equivalent to performing a sensitivity analysis with Zernike coefficients. In this example, the sensitivities to Zernike coma coefficient is then targeted to be equal in the merit function. Figure 59 shows the optimized design.

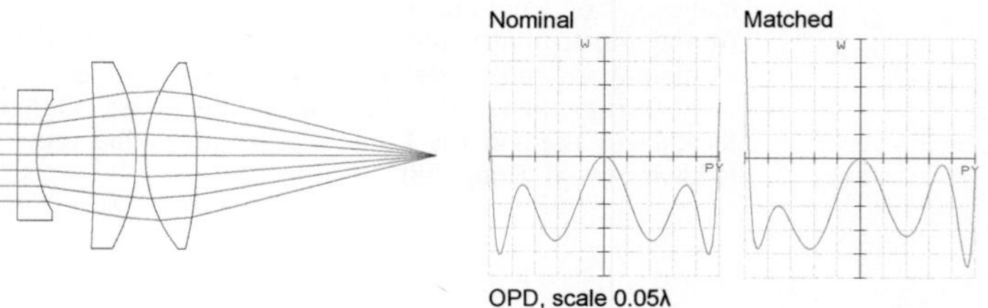

Figure 59
Laser focusing lens
optimized for
matching tolerances
with Zernike
approach

In contrast to the multi-configuration approach, this method results in almost perfect compensation. The remaining difference between the wavefronts results from higher order errors that were not included in the optimization. However, this better match comes at the price of a reduced nominal performance. Still, diffraction limited performance is achieved.

If compensation can be reduced to specific aberrations, symmetric or asymmetric aberration coefficients can also be used. This is similar to using Zernike coefficients but different in that the effects of specific surfaces are taken into consideration. In this example vector aberration theory can be employed to achieve compensation of third-order coma. The calculation can be implemented as a macro or directly in the merit function.

Recognizing that the ratio of the chief ray angle of incidence of the optical axis ray and the reference ray determines the shift of a surface's coma contribution, this is relatively straight forward. For small deviations the real angle may be used instead if the paraxial one if the ray-tracing software does not permit disturbed paraxial ray-traces. Using a dummy configuration (CONF 2 in Figure 60) with perturbed components that does not enter the imaging properties but is only used to determine the optical axis ray the formulation can be expressed as follows:

Figure 60
Merit function formulation (Zemax notation) to reduce third-order coma effects

	Op#1	Op#2	Hx	Hy	Px	Py		Value
1: CONF	2							
2: REAY	4	1	0.0000	0.0000	0.0000	0.0000		−0.0200
3: CVVA	4							0.0836
4: PROD	2	3						−1.6722E−003
5: RAGB	3	1	0.0000	0.0000	0.0000	0.0000		0.0000
6: SUMM	4	5						−1.6722E−003
7: PARY	4	1	0.0000	1.0000	0.0000	0.0000		0.0556
8: PROD	3	7						4.6498E−003
9: PARB	3	1	0.0000	1.0000	0.0000	0.0000		5.9893E−003
10: SUMM	8	9						0.0106
11: DIVI	6	10						−0.1572
12: COMA	4	1						−7.1366
13: PROD	11	12						1.1217

Using macros is always more time consuming and may impact convergence, while a definition in the merit function shows the best convergence of the three methods presented. This is probably due to the small number of additional calculations. The result of the optimization is shown in Figure 61.

Figure 61
Laser focusing lens
optimized for
matching tolerances
with third-order
vector aberration

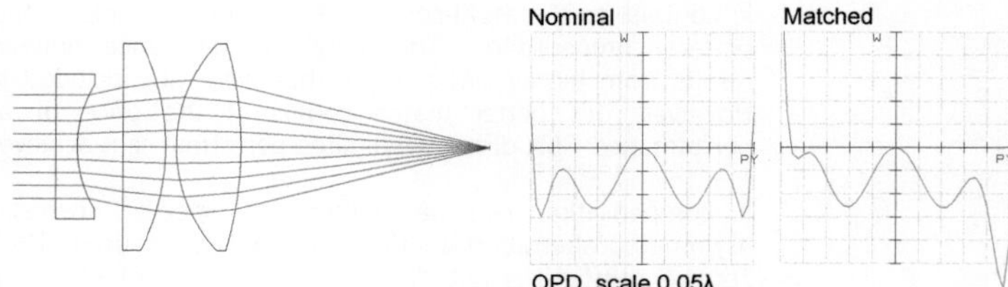

On a scale of 0.05λ the OPD seems to show coma and tilt effects for the matched tolerances. This is indeed true, but the coma is of higher order while third-order coma is reduced to zero as desired. This can be best shown by looking at the saggital ray fan plotted with y-aberration. Third-order coma would produce a parabolic shape in PX, while in this example, the polynomial order is higher.

Figure 62
Ray-fans showing
deviations in the
meridional plane
(EY) for meridional
(PY) and saggital
pupil variation (PX).
Higher order coma
can be observed for
the plot of EY over
PX.

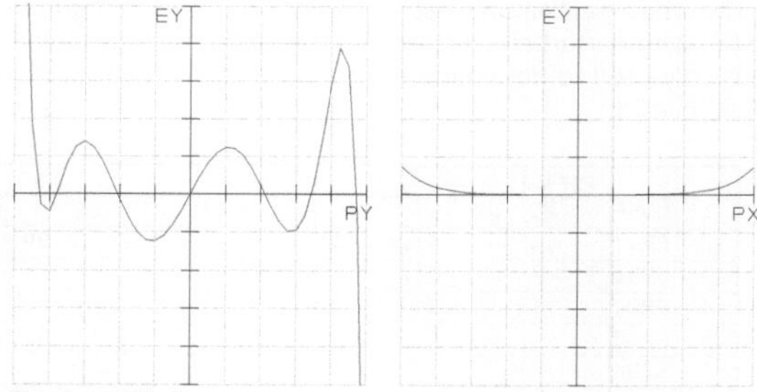

It is also interesting to observe changes of the sensitivity of lens 2 while optimizing the match of lens 1 and 3. The sensitivity of lens 2 constitutes a secondary tolerance and should therefore have a low sensitivity. Through adaption of the matching range the sensitivity may however increase. In this example the following results are obtained (Figure 63), showing that the sensitivity of lens 2 with respect to decenter has decreased as well. However, this cannot generally be expected and strategies to simultaneously desensitized secondary tolerances while adapting matching ranges need to be employed.

Type of error		Design adaptation method		
		Multi-Config	Zernike	Vectorial coma
Lens 1	wavefront tilt	0.652	0.763	0.781
Lens 1	coma	0.248	0.291	0.297
Lens 2	wavefront tilt	-0.018	0.023	-0.006
Lens 2	coma	-0.001	0.019	0.011
Lens 3	wavefront tilt	-0.636	-0.791	-0.781
Lens 3	coma	-0.247	-0.310	-0.309

Nominal Strehl	0.998	0.996	0.993

If more than two parameters can contribute to a matching the number of possible tolerance combinations increases (e.g. + + -, + - +, etc.) and testing every combination with the multi-configuration approach would be very inconvenient. Using Zernike coefficients or aberration coefficients has then clear advantages, as their sign indicates possible choices. The situation further complicates if multiple specifications are to be compensated.

5.2.2 Desensitization of secondary tolerances

Desensitization has its origins in the reduction of sensitivities to centering and tilt tolerances as these are frequently producing the worst imaging errors. On-axis coma arising from decentered components and surfaces is particularly undesired. Even though the methods reviewed in section 2.3.4 are not in common use, they can be regarded state of the art. The important question with respect to desensitization of optical systems for combinatorial assembly is, which methods are most suitable, particularly in conjunction with matching adaption. Most important criteria for the application to optimize matching behavior are convergence, quality of the solution, amount of restriction of the solution and effort to set up the criteria.

Computational effort and a limited (by the simulation software) number of possible configurations as well as complicated merit functions render multi-configuration approaches unsuitable for a large number of parameters. Methods with statistically distributed configurations to reduce this dependency may also converge better, but can result in poorly centered designs as the nominal solution is no longer in the 'center' of the perturbations. In contrast to this brute-force approach Isshiki's method of reducing angles of incidence is usually faster, converges better and shows better exploration of the design space. However, reducing ray angles does not account for the real dependency of sensitivity to centration errors.

The following example illustrates the effects desensitization can have in combination with combinatorial assembly. Designed to operate at two distinct wavelengths, 1.064 nm and 532 nm, the following laser scanning lens is fully color-corrected and diffraction limited over the entire field, enabling accurate processing as well as process control.

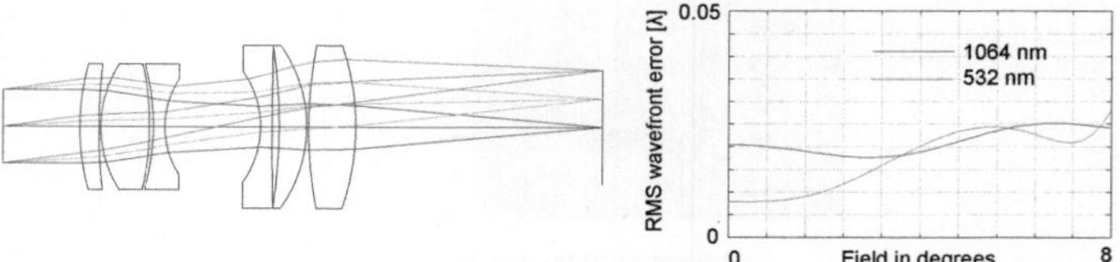

Figure 64
Color corrected laser scanning lens with 20 mm Aperture, 16° full field, focal length 110 mm

A sensitivity analysis reveals that lenses two, three and four are most sensitive to assembly tolerances. Correcting axial and lateral color is to a certain extent achieved by the split achromat in the first group and this is the most critical element. Decentration and tilt of the components result primarily in coma and astigmatism on-axis as well as in the field. Centration tolerances of the individual lenses need to be less than 10 μm to remain diffraction limited.

In order to investigate compensation of these errors, field points in every coordinate direction need to be considered and the effects of coma and astigmatism evaluated for both wavelengths. The total number of relations is thus comparably large and compensation can be expected to be very complicated. Realizing that the split achromat is most sensitive, it turns out that bringing these two lenses into alignment already results in very good compensation. Only lens four then has a significant impact on performance and is an ideal candidate for desensitization if lenses two and three are treated with combinatorial assembly.

Of the above mentioned methods, multi-configuration desensitization and reducing ray angles as proposed by Isshiki did not succeed. However, in reducing the coma contribution of the front surface of lens four, its sensitivity to assembly tolerances can be reduced. The desensitized design is shown in Figure 65.

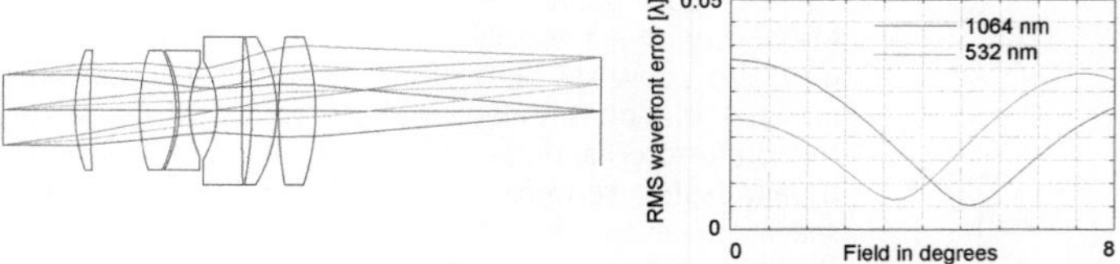

Figure 65
Color corrected laser scanning lens designed to be combintorially assembled

From the RMS wavefront curve it can be concluded that nominal performance is somewhat reduced. In fact, a certain amount of field curvature developed and is balanced against defocus such that a point at half field performs best.

If the two designs are compared for a 50 µm decenter of lens four, the following point spread functions can be determined (Figure 66). The Strehl ratio of the decentered lens remains above 0.84, while the original lens exhibits a drop to 0.68 showing coma and astigmatism.

Figure 66
Sensitivity to 50 µm
decentration of lens
four

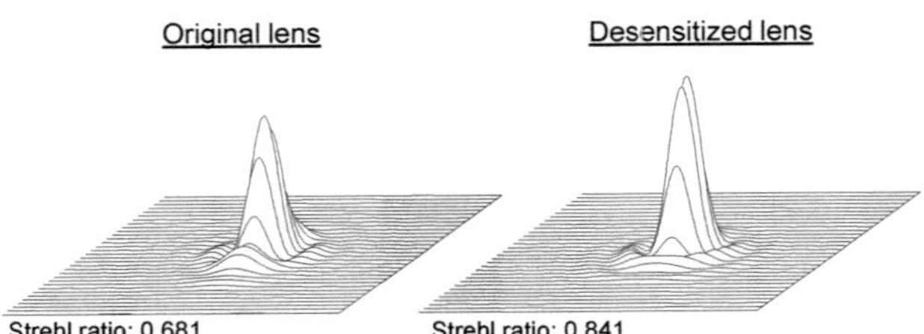

In changing the design, the sensitivity of the split achromat increases only slightly. If through combinatorial assembly the *difference* in decentration of these two components can be reduced to less than 14 µm, diffraction limited performance can be expected. As illustrated in chapter 4, a reduction of a factor of four in the tolerances can be achieved for comparably small quantities such that a decentration tolerance of 50 µm seems possible. The entire lens can then be assembled in a Poker-chip assembly, where every element is mounted in a cell. This allows for very easy assembly and only the centrations of lens two and three need to be characterized and suitable lenses selected for combinatorial assembly. The remaining components can be randomly chosen.

Further improvements of desensitization could be achieved if vector aberration theory is employed to define a measure of sensitivity for different errors such as third-order coma. Using aberration theory gives a much better representation of the actual sensitivity than ray-angles or symmetrical aberration coefficients alone.

Let again denote i_{OAR} the angle of incidence of the optical axis ray, i the chief ray angle and w_{131} the ordinary coma contribution. As mentioned before, the surface contribution a_{131} to third-order coma is given as:

5.3

$$a_{131} = \frac{i_{OAR,j}}{\bar{i}_j} \cdot w_{131}$$

If this is used as a sensitivity measure, two effects need to be distinguished. First, the effect from disturbing a single surface and second, effects that are induced by other surfaces due to the perturbation. This can be expressed as a differential and the exact calculations can be found in appendix 1.1.

In contrast to desensitization of centering tolerances the sensitivity of lens parameters (e.g. surface form, thickness of material properties) is rarely considered a variable. Variations of curvatures, thicknesses, material are often uncritical. However, air spaces and ellipticity of surfaces are frequently critical. Similar approaches can be employed to desensitize these parameters.

5.3   Global optimization strategies

So far, changes of the optical design were induced by altering an existing design form. While the techniques described above can find solutions which differ substantially from the original design it should be possible to go even further and consider global optimization.

The optical merit function has a multi-dimensional topography, the design landscape. As stated above, the existence of long and narrow valleys seems commonly accepted and is well documented for simpler systems (e.g achromatic doublet). Understanding the nature of the merit function can be critical in finding different minima and is certainly necessary if it is attempted to find them all. Most optimization methods – including Damped Least Squares (DLS) – have in common, that they can only locate local optima because the number of parameters and nonlinearity of the merit function make optimization a difficult task. Global optimization has been first investigated in other disciplines and adapted in optics. Of the different methods, Bociort's Saddle Point method seems well suited to systematically investigate the different solutions of a design. Bociort pictures the design landscape as hills and valleys and provides a method to generate saddle points which can be used to locate new minima. It is then possible to systematically search the merit function for solutions and find every local optimum [VTUR09]. This method shall be applied to the laser focusing lens example comparing the properties of different solutions with respect to combinatorial assembly.

The saddle point method works as follows: Starting from a single optimized lens a so called null-lens is created. This null lens is a shell of zero thickness on one surface of a lens element with identical curvatures and material like the single lens. This creates a saddle point in the higher dimensional parameter space because the derivative of the merit function of the original parameters remains the same and the derivative with respect to the new variables is zero. From there, two start configurations can be created, one with slightly smaller curvatures of the null lens and one with slightly larger curvatures. Optimization will then find two optima on either side of the saddle and thickness and air space can gradually be increased to create a useful solution [VTUR09]. The theory can be generalized and arbitrary null lenses placed anywhere in the system as long as a saddle is constructed. That is, the derivatives of the new variables are zero.

The systematic can be repeated to generate (almost) all the solutions of the laser focusing lens and create a so called network. The network consists of optimized solutions and the saddle points between them. Following the described procedure the following five solutions were derived and their differences with respect to combinatorial assembly discussed (Figure 67).

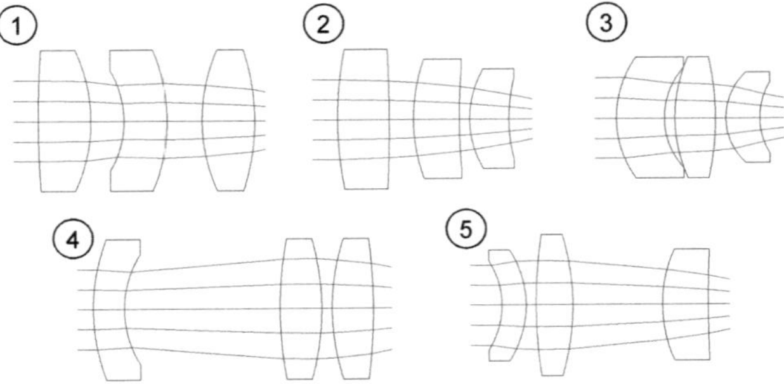

Figure 67
Different local minima for the laser focusing lens example, 1° field, 8 mm entrance pupil

Analyzing the changes for Zernike coefficients of tilt and coma delivers the following results for a 20 µm decenter and 50 millidegree tilt of the three lenses.

Figure 68
Change table of Zernike coefficients for tilt and coma

		System 1		System 2		System 3		System 4		System 5	
		Z3	Z7	Z3	Z7	Z3	Z7	Z3	Z7	Z3	Z7
DEC	L1	-0,133	-0,067	-0,013	-0,006	0,192	0,102	0,240	0,126	0,123	0,067
TILT	L1	-0,053	-0,027	-0,001	0,000	-0,096	-0,051	-0,106	-0,056	0,042	0,023
DEC	L2	0,288	0,151	-0,008	-0,004	-0,068	-0,033	-0,152	-0,078	-0,111	-0,057
TILT	L2	0,089	0,047	-0,004	-0,002	-0,004	-0,002	-0,007	-0,003	-0,034	-0,017
DEC	L3	-0,159	-0,082	0,007	0,004	-0,124	-0,064	-0,093	-0,048	-0,016	-0,008
TILT	L3	-0,046	-0,024	-0,007	-0,004	0,031	0,016	-0,063	-0,032	-0,024	-0,012

Nom. Strehl	0.975	0.817	0.976	0.982	0.923

The results show an extreme insensitivity of design two, as expected, because it has the least bending of rays. The design form is often used for test-lenses where high performance is required, typically with even more lens elements. Hence, this type of design would often be a good choice for classical assembly. Design two has, however, the worst nominal performance.

Design five shows very similar sensitivities for element one and two, while element three is comparably insensitive. This is ideal for combinatorial assembly. Very similar tolerances can be assigned for tilt and decenter and – if necessary – the methods described above can be employed to further adapt tolerance regions. It is in fact possible to design element three of this solution to be aplanatic, the first surface being concentric and the second aplanatic. Figure 69 shows the results of a tolerance analysis for design five.

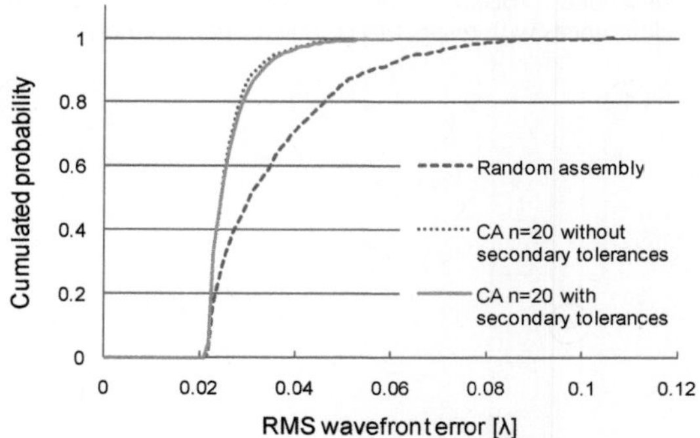

Figure 69
Simulated
performance after
combinatorial
assembly of design
five

In contrast to the tolerance analysis results of the laser focusing lens that was shown at the end of chapter 4 (Figure 52), the effects of secondary tolerances are significantly reduced for the depicted solution. Hence, without specifically optimizing a lens, local optima of the design landscape may be suitable for the application of combinatorial assembly. Selecting a good start design as the one just found can help to further increase the compensation potential.

5.4 Summary and conclusion

Combinatorial assembly can increase the performance of optical systems and reduce their cost. But it works only if the selected design is such that parameter deviations can compensate each other, such that secondary parameters have little influence and such that tolerances match in a cost-efficient way.

This chapter shows that there are indeed certain lens design alternatives that are better suited for combinatorial assembly than others. This is exemplarily shown for a three element focusing lens and methods to force a design to suitable solutions during optimization are discussed. Using multiple design configurations it is possible to increase the match of tolerances and optimize it to minimize the cost of tolerances.

Optimizing the lens design for combinatorial assembly has multiple positive effects. First, the performance increase through combinatorial assembly is maximized because tolerances can better match. Second, secondary parameters can be more easily desensitized which in turn reduces additional errors. Third, the nominal performance can be higher than for a fully desensitized system. Together, optical systems with higher nominal performance and reduced tolerance requirements appear possible.

The next chapter presents simulated and experimentally verified test cases.

6   Examples of combinatorial assembly

Differentiating combinatorial assembly in compensation of performance criteria of different order (paraxial deviations, third-order aberrations and higher order aberrations) on the one hand and in compensation of symmetrical and asymmetrical aberrations on the other, spans a wide field of applications. In addition, cost reduction, size and weight reduction, avoidance of additional mounts, adjustment procedures and mechanical complexity provide good reasons for the application of combinatorial assembly. The following examples represent case studies exemplarily representing the different dimensions.

Figure 70
Classification of
combinatorial
assembly examples

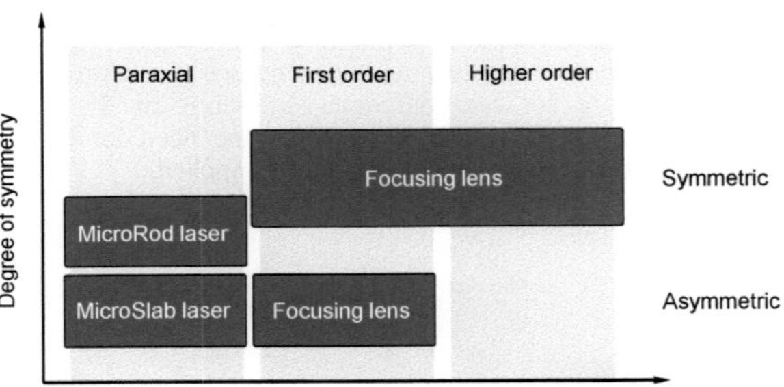

6.1   Planar assembly of solid state lasers

With planar assembly of optical components and small solid state lasers, a technology substituting manual assembly has recently been pushed forward. Basic idea of the concept is the placement of optical and electronic components on a flat substrate, similar to surface mounting in electronics (SMD-Technology); adjustable mounts are completely omitted. A flexible and quick assembly of individualized laser systems is aspired and two similar systems emerged from the Cluster of Excellence "Integrative Production Technology for High-wage Countries" at RWTH Aachen University: A miniaturized, direct diode-pumped slab laser (MicroSlab) and a rod laser (MicroRod). Both systems are developed for marking applications and deliver nanosecond pulses at an average power of a few Watts. Figure 71 depicts the planar setup of MicroSlab and MicroRod.

Figure 71
MicroSlab base
plate (left) and
MicroRod (right)

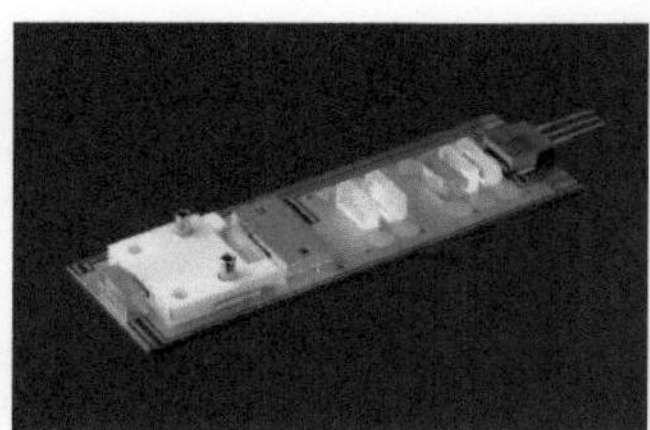

The size of both systems is about 100 mm in length, demonstrating that a very compact module can be realized using the planar assembly technology.

6.1.1 Centering errors of MicroSlab's pump optics

Planar assembly implies that centering tolerances impact the relative position of component's axis of symmetry with respect to the base surface. This misalignment results in an undesired deviation of the path of light as shown in Figure 72 for two different centrations of the laser diode. In order to compensate this deviation without the need for classical iterative alignment procedures combinatorial assembly is applied.

Figure 72
Meridional view of
MicroSlab's pump
optic (aperture 4x
exaggerated) for
centered and
decentered diode

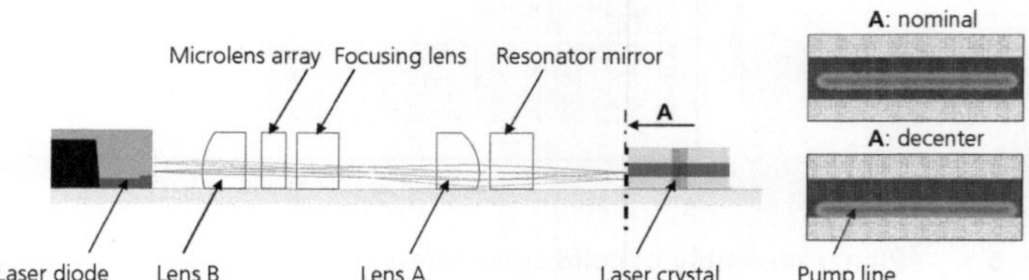

Before the light enters the laser crystal, the laser light emitted by the diode is homogenized in one direction by a micro-lens array and demagnified in the other, adapting the size of the laser beam. Imaging the diode laser onto the laser crystal's facet achieves the desired reduction in beam height. Principally, a single lens is sufficient to fulfill this task.

A better reproduction of the beam profile can be achieved using a double-sided telecentric lens. For the centering error under consideration, the vertical positions of laser diode, crystal and lenses A and B are of relevance. In addition, not only centration but also perpendicular incidence of the beam is of importance. The arrangement has just enough degrees of freedom to adjust both centering and angle of incidence.

The working principle of solid state lasers inevitably leads to a thermal gradient inside the laser crystal resulting in a gradient of the index of refraction that acts like a lens. For perfectly symmetrical incidence of the pump beam the thermal lens will also be perfectly symmetrical. Asymmetric thermal lenses however lead to degraded beam quality as well as a change in beam direction and therefore need to be avoided. Particularly, the relative position of the incident pump light with respect to the laser medium is important due to thermal lensing [FUNC10] Combinatorial assembly can reduce the centering error without the need for additional mounts such that planar assembly can be used. This type of correction classifies as a first order compensation using asymmetric parameter deviations. Aberrations do not play a significant role.

From sensitivity analyses, the dependence of the pump light position to the center of the crystal Δy can be related to the critical parameters. These are the height of the laser diode's emission ΔD, the height of the crystal's center ΔC as well as decentrations ΔA and ΔB and tilts φ_A and φ_B of the lenses. The linear model (centrations in mm and angles in degrees) then reads as:

6.1
$$\Delta y = -0.744\Delta_D + \Delta_C - 1.43\Delta_A - 0.310\Delta_B + 0.103\varphi_a + 0.043\varphi_B$$

Combinatorial assembly starts with a planar substrate and laser diode and crystal pre-assembled. Then, this module is characterized to determine ΔD and ΔC before lenses are selected for combinatorial assembly.

Lens tolerances have been determined in advance using digital image processing. The cross-sections of the cylindrical lenses are recorded with a telecentric machine vision camera and these images evaluated with the help of edge detection. The uncertainty of lens measurements has been determined to be a few microns and the tolerance distributions of 50 lenses provide a good estimate for tolerancing (see appendix 1.4). Once all the values are recorded, they are fed into the database system and the optimal matching is found and then assembled.

Based on the measured tolerance distributions an analysis of combinatorial assembly is performed for a production of ten systems and compared to regular assembly (Figure 73). As an additional uncertainty, measurement errors as well as a variation in lens decenter due to the joining process have been included in the analysis. The results show that combinatorial assembly can reduce the centering error of the worst system in a set of ten by a factor of four.

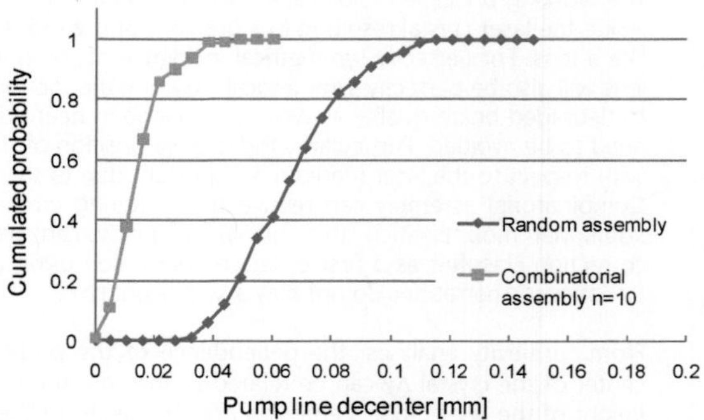

Experimental verification of these results requires the collection of statistical data and hence, many repetitions of assembling a series of *n* systems. A few hundred to a few thousand systems would be necessary. Alternatively, the performance predictions of the linear model can be tested. If the performance of single systems is accurately predicted by the model, repetition of the computer experiment is legitimate. Verification of the model is thus performed through comparisons of simulated centering and experimentally determined centering, based on the same parameter measurements. The resulting pump beam position on the laser crystal is determined by observing the emitted fluorescence. Figure 74 shows experimental and simulated pump light positions.

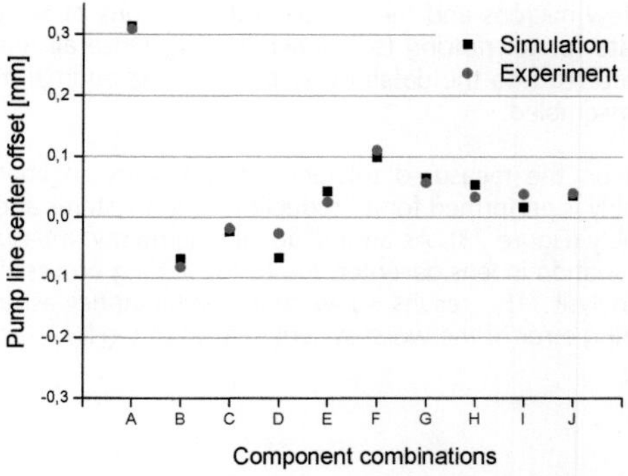

The observed mismatch includes uncertainties of the parameter measurements, the measurement method to determine the position of the pump light as well as joining tolerances. Generally, the agreement is better than 20 µm except for combination D. Here, the deviation could be traced back to a bad metal coating (required to facilitate the joining process) at the bottom surface of one of the lenses which disturbs edge detection.

6.1.2 Symmetry aspects of MicroRod

In analogy to the MicroSlab laser, centering errors of MicroRod causing a deviation of the pump light are compensated with combinatorial assembly. Using symmetries it is however possible to realize a significant increase in centering precision even for single systems.

Flexibility in the design of the pump lens permits the use of two symmetrically arranged axial-symmetric lenses (Figure 75). In order to facilitate planar assembly, the lenses are ground to a quadratic shape resulting in four possible positions as the lenses can be assembled on any side. A total of 32 combinations of which 16 are distinctly different are thus feasible for a single system and the most suitable can be selected. In specifically designing a symmetrical system and choosing a suitable separation between the components it is thus possible to increase the possibilities of combinatorial assembly.

Figure 75
MicroRod pump lens
assembly

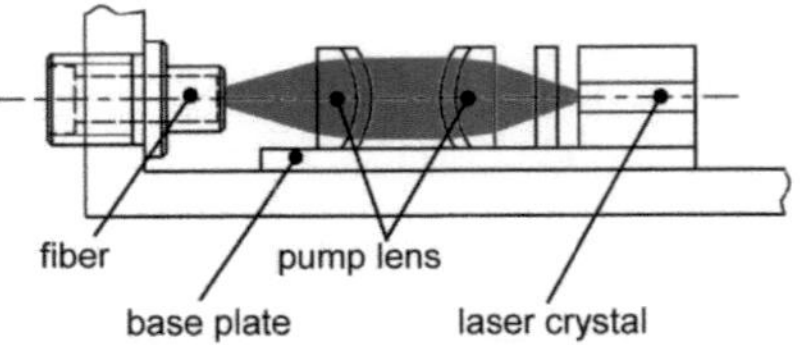

As with the MicroSlab, characterization of the centrations of pump light, laser crystal and lenses is performed using digital image processing. Because the lenses are aspherical it is significantly more complicated to determine lens centrations from the acquired images of cross sections.

The comparably large deviations of the estimated center position of the laser crystal cannot be directly compensated by the natural deviation of centering errors of the lenses as their effect is too small and their tolerances would have to be larger than the standard tolerance. In order to adapt the usable tolerance range the lenses are manufactured with a deliberate centering offset. Due to the symmetry of the lenses this artificially enlarges the tolerance range and can also be used to moderately change the distribution. Tolerances are given to the lens height and relative centration (10 µm + 20 µm).

Simulations of the quality improvement through application of combinatorial assembly predict a reduction of the centering error from 60 µm to 20 µm for a single assembly and a 90% yield (Figure 76).

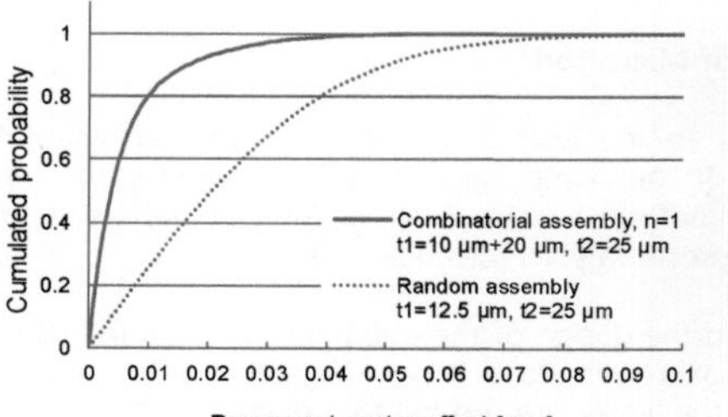

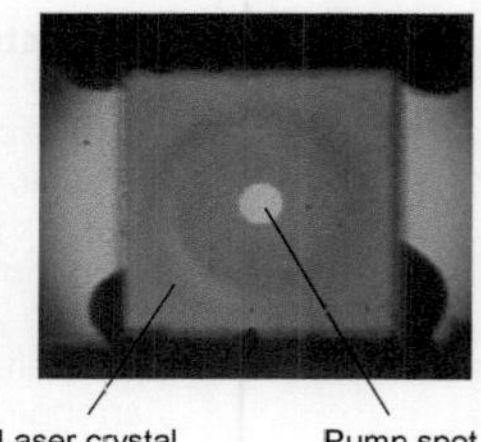

Figure 76
Simulated deviation for n=1 and measured assembly tolerances (left): t_1: lens centration, t_2: crystal position camera image of assembled system (right)

The final assembly exhibits a measured deviation of 20 µm and Figure 76 shows the nicely centered pump beam position on the crystal's front facet after assembly. While the vertical centration is determined by the lens combination, the horizontal centration can be optimized through alignment as planar assembly does not restrict this movement.

6.2   Laser focusing lens

Focusing lenses have two major applications in laser technology. Focusing single mode lasers to achieve minimal spot sizes and imaging the light from an already collimated fiber (top-hat focus). Increasingly, process control through focusing lenses and processing at multiple wavelengths is performed resulting in additional demands with respect to color correction and resolution at medium sized fields. Very characteristic for focusing lenses is the dominance of spherical aberration that requires correction. Field aberrations do not play a significant role due to the limited size of the field. Two aspects are of particular importance:

1. For focusing lenses with numerical apertures above ~0.1 spherical aberrations of third- and higher orders need to be balanced in order to correct the lens.

2. Correction of field-dependent aberrations is not required for axial symmetry, however, decenter and tilt can produce large amounts of coma. This is especially true if the system is not corrected for coma initially.

Both cases have potential for the application of combinatorial assembly; to maintain sensitive aberration equilibria and for the compensation of axial coma.

6.2.1 Compensating aberration equilibria

Originally designed for laser coating processes, the following lens (Figure 77) with an extremely large numerical aperture of 0.6 is very versatile. Aberration correction is achieved by balancing spherical aberration of multiple orders and diffraction limited performance is achieved on axis. In addition, the lens can be used for a wide range of different wavelengths from 1 µm to 2 µm. The lens is not truly color corrected but can be adjusted for any wavelength in the operating range by moving the negative lens in order to correct spherochromatism, the change of spherical aberration with wavelength (Figure 77). Moving the lens requires additional mounts, adds complexity, weight and proper centration needs to be maintained during the movement which is complicated by the fact that the negative lens is the most sensitive of all to centration errors.

Figure 77
Focusing lens (left) and in-focus OPD for two wavelengths showing variation of spherical aberration (right)

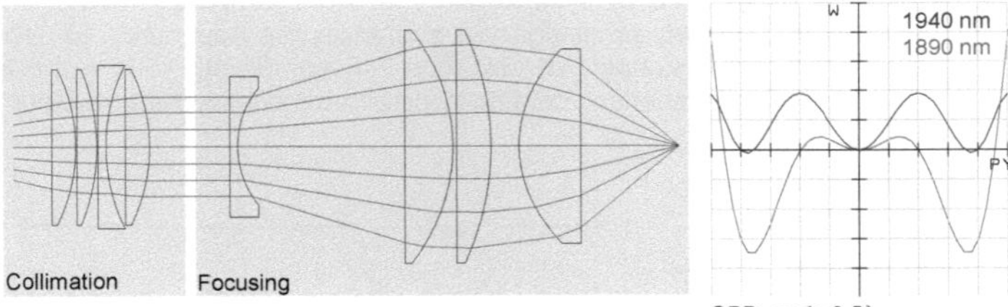

Even if this lens were to be used at a single wavelength, adjusting the position of the negative lens is required to correct spherochromatism as shown in Figure 77. This would be very costly in a series production and a tolerance analysis reveals that it is not possible to remain diffraction limited without adjustment. Figure 78 shows the change table for RMS wave aberration

Figure 78
Change table derived from a sensitivity analysis of curvature and thickness tolerances of the focusing lens (lens data in appendix1.5)

Tol	Surface	RMS wave	ZERN4	ZERN9	ZERN16	ZERN25
TTHI	11	0.0303	-0.0014	-0.1139	-0.0120	-0.0019
CURV	11	0.0349	-0.0016	-0.1271	-0.0076	-0.0019
TTHI	12	0.0166	-0.0010	-0.0765	-0.0077	-0.0012
CURV	12	0.0042	0.0005	0.0336	0.0031	0.0007
TTHI	13	0.0003	-0.0001	-0.0071	-0.0017	-0.0003
CURV	13	0.0626	0.0026	0.1931	0.0106	0.0023
TTHI	14	0.0001	0.0000	-0.0050	-0.0012	-0.0002
CURV	14	0.0024	0.0003	0.0241	0.0018	0.0003
TTHI	15	0.0004	0.0001	0.0072	0.0015	0.0000
CURV	15	0.0343	0.0016	0.1243	0.0083	0.0017
TTHI	16	0.0067	0.0006	0.0436	-0.0006	0.0005
CURV	16	0.0724	-0.0029	-0.2146	0.0344	-0.0062
CURV	17	0.0205	-0.0011	-0.0879	-0.0074	-0.0015

From the tolerance analysis it becomes obvious that spherical aberration of third-order is sufficient to describe the effects due to parameter perturbations. Fortunately, higher order aberrations do not vary much. As a result, minimizing the effects of third-order (ZERN 9) aberration will guarantee a maintained aberration balance. Hence, only two modules are required during combinatorial assembly. Module composition is mainly driven by mechanical restrictions and measurement capabilities. Grouping the first two lenses and the last two lenses allows for comparably easy testing, as the wavefront between the two groups is almost plane. Measuring complete modules instead of single tolerance deviations is very suitable for single specifications and can be easily realized with an interferometer or wavefront sensor. The measured Zernike coefficients can directly be used to sort and match modules.

Simulations of combinatorial assembly reveal a possible reduction of the errors induced by symmetrical tolerances. Figure 79 shows results of the tolerance analysis simulation (for tolerances see Appendix 1.5). From these it is obvious that strategic combination can significantly reduce the error, while statistical combination could have increased the wavefront deviation considerably.

Figure 79
RMS wavefront error for combinatorial assembly and random assembly of the focusing lens

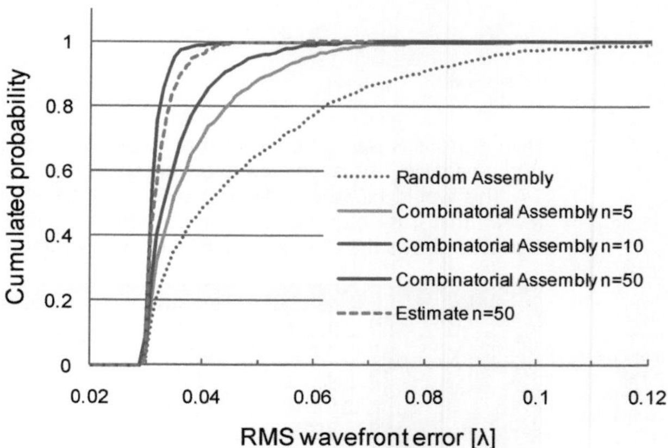

In this example, adjusting the back focus for optimal RMS errror has been included in the analysis. Note that the nominal system has an RMS wavefront error of about 0.03. In addition, the estimated performance according to the scaling model is shown. The curve for n=50 is approximated from data for n=10 and illustrates that the approximation is not exactly valid in this case. This is due to the nonlinear behavior of the performance function. Still, the approximation gives a good indication and errs on the conservative side.

6.2.2 Compensating on-axis coma

As mentioned before, the focusing lens (Figure 77) is not only sensitive to spherical aberration and spherochromatism, but also to centration. Decentered and tilted components are one of the main causes for inferior imaging performance of optical systems. Since the symmetry of a lens is broken if components are off-center or tilted, optical path lengths change and field aberrations such as coma appear on axis.

As demonstrated before, vector aberration theory provides insight into the mechanism involved and motivates the possibility to reduce such effects by combining multiple components with matching errors. As decentration and tilt due to tolerance are relatively small, third-order vector aberration theory is frequently sufficient to describe the effects. In many cases decentration problems can even be reduced to on-axis coma.

Figure 80 shows the sensitivities for lens tilt and decenter and it can be seen that the asphere (Lens 4) has already been designed to be insensitive against decenter.

Figure 80
Change table of
opto-mechanical
sensitivities of the
focusing lens

Tol	Lens	RMS wave	Strehl	Zernike 3	Zernike 8
DEC	1	0,232	-0,752	1,411	0,757
TILT	1	0,064	-0,248	-0,508	-0,289
DEC	2	0,119	-0,506	-0,873	-0,446
TILT	2	0,002	0,003	0,047	0,042
DEC	3	0,074	-0,303	-0,610	-0,319
TILT	3	0,013	-0,040	-0,218	-0,115
DEC	4	0,000	0,007	0,072	-0,006
TILT	4	0,111	-0,462	-0,800	-0,425

The sensitivity analysis provides insight into the types and magnitudes of errors. In order to compensate on-axis coma it is of importance to realize that the error is a vector quantity. A decentered or tilted component is characterized by the magnitude of the error and its direction. If modules can be combined in discrete or continuous angular positions (similar to clocking) the positive effect of compensation through combinatorial assembly can be enhanced. Figure 81 shows the results. For a limit of 0.07 RMS wave (diffraction limit) the yield can be improved from 60% to over 90%.

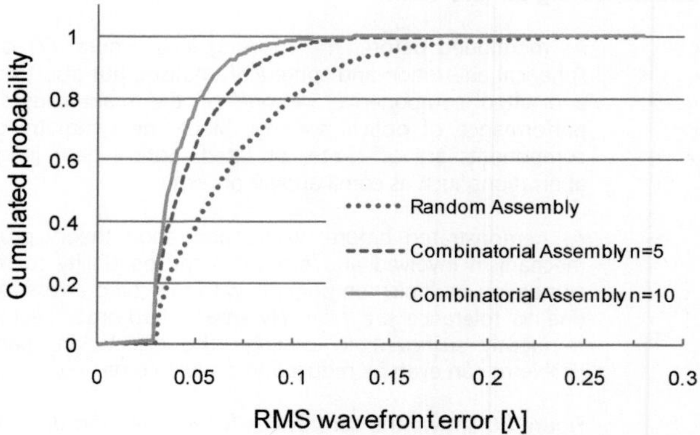

Figure 81
Tolerance analysis
results for
combinatorial
assembly of the
focusing lens
reducing axial coma

If the problem can be reduced to on-axis coma, every module can be characterized as a whole and it is not necessary to characterize every particular parameter of significant sensitivity. The number of measurements can thus be comparably small.

As we have seen, spherical aberration needs to be corrected as well, but depends on a different set of parameters. In this case it is still possible to measure combined effects and if three modules are defined both specifications can be met simultaneously.

7   Conclusion and Outlook

Producing high quality optical systems at reasonable cost is a challenge. Manufacturing and assembly tolerances have a large impact on both performance and cost, and frequently compensation is required to achieve the demanded performance. Alignment, the most common compensation strategy requires an iterative approach during assembly, taking time and skill, while selective assembly is only suitable for mass production.

In this dissertation, combinatorial assembly is presented as an alternative compensation strategy which does not require iterations and is suitable for automated small series production. A pool of components and subassemblies, necessary for the assembly of a series of systems, is characterized prior to the assembly and then the best system configurations are selected. The selection builds on measurement results which are stored in a database before the configurations are optimized with the help of a ray-tracing simulation model. In order to facilitate this approach, an integrated software tool which automatically suggests the optimal assembly strategy is realized for the first time.

In contrast to selective assembly, combinatorial assembly uses exact measurement results instead of classified components. Mismatch of tolerance classes resulting in unused components is therefore avoided and the possible degree of compensation for small series production is enhanced. This is practically demonstrated for the example of micro lasers, where combinatorial assembly enables automated planar assembly. With accurate measurements a significant performance increase is determined even for small and single series production. In addition, multidimensional problems with interrelated specifications can be treated. The investigations show that the degree of compensation is a function of the production volume and a simple scaling law is given to predict the performance increase for changed production volumes. In addition, the final performance of optical systems is influenced by secondary parameters; component attributes which were not measured prior to the assembly.

It is pointed out that the application of combinatorial assembly to optical systems is significantly influenced by the existence of numerous interrelated performance specifications. This requires a delicate choice of matching parameters, modules and tolerances by the optical engineer before components are manufactured. It is otherwise very unlikely that applying the method afterwards to already manufactured components will increase performance. In order to identify useful modules and subassemblies for combinatorial assembly a compensation model taking the discrete nature of combinatorial compensation into account is derived, fostering the understanding of combinatorial assembly. Imaging performance is particularly complex because virtually every system parameter has an influence. It is shown that in order to reduce the complexity of selecting matching parameters, sub-specifications such as Zernike polynomials or aberration coefficients need to be considered.

Predicting the as-built performance of combinatorially assembled systems with high accuracy necessitates a dedicated tolerance analysis. While preliminary selection of an assembly strategy can be based on linear approximations, the effects of discrete compensation, secondary tolerances and production volume need to be analysed for realistic performance measures. A concept for such analyses is proposed. It consists of a Monte Carlo approach to generate random parts and components and finds the optimal system configurations before secondary tolerance effects are analyzed. The concept is universally applicable to problems that can be modelled with ray-tracing and is implemented using a combination of ray-tracing software, logic calculator and database. For the first time ever, it is possible to accurately analyse the impact of matching tolerances, production volume and secondary tolerances on the performance of combinatorially assembled optical systems.

Using the tolerance analysis tool it is shown that in many cases performance can be increased by a factor of four to ten for production volumes between 10 and 50 systems. For the assignment of tolerances, an effective sensitivity is introduced and existing tolerancing models modified to assign cost-optimal tolerances. It is presented that the effective sensitivity depends on the number of combined systems and can be derived from tolerance analyses with the developed analysis tool. For optimal compensation, tolerance distributions should match each other and it is illustrated that this can be difficult to realize for some lens designs due to manufacturing limits. As a result, cost-optimal tolerances cannot always be assigned or the compensation effect is reduced.

In order to alleviate this restriction, design strategies increasing combinatorial compensation are derived. Manipulating the optical design from the outset to suit combinatorial assembly can shift tolerance sensitivities from one component to another. Compensation can be enhanced and the influence of secondary tolerances reduced. Different concepts are developed and exemplarily proven to increase the applicability of combinatorial assembly in achieving a better match of tolerances and a reduced sensitivity to secondary tolerances. In using combinatorial assembly in conjunction with desensitization, systems with higher nominal performance yet reduced tolerance demands can be build. This is an entirely new approach and a first step towards a more integrated development of optical systems.

Further improvements can be made in many areas. The speed of optimization and tolerance analysis can be enhanced by using differential ray-tracing and new optimization algorithms, while the inclusion of tolerance cost-models can improve cost-optimal tolerancing. In addition, combinatorial assembly is only one of many assembly methods and by no means the solution for every problem. Depending on the application, other strategies may be more suitable and the development of high-performance optical systems will benefit most from a flexible or even dynamic choice. An intelligent automated assembly system will select the right strategy at the right time and first steps have been taken in employing self-optimizing concepts based on multi-agent software [LOOS11]. Further advances in integrative design, assembly and tolerancing seem possible. Most importantly, the communication between engineers of different technical disciplines involved in the development process needs to be enhanced. Interactive development platforms which connect lens designers, mechanical engineers and production specialists are required to synergetically derive optical systems of highest performance.

II Bibliography

[ADAM87] G.P. Adams, "Tolerancing of optical systems," University of London, Imperial College of Science and Technology, 1987.

[AHMA97] A. Ahmad, Handbook of optomechanical engineering, CRC Press, 1997.

[ANDE80] T.B. Andersen, "Automatic computation of optical aberration coefficients," Applied Optics, vol. 19, Nov. 1980, p. 3800.

[BASS09] M. Bass, V.N. Mahajan, E.V. Stryland, and O.S. of America, Handbook of optics, Volume II, McGraw Hill Professional, 2009.

[BECK05] E. Beckert, "Ebene Keramiksubstrate und neue Montagetechnologien zum Aufbau hybrid-optischer Systeme," Technische Universität Ilmenau, 2005.

[BLIE08] J. Bliedtner and G. Gräfe, Optiktechnologie, München: Fachbuchverlag Leipzig im Carl Hanser Verlag, 2008.

[BOCI05] F. Bociort, "Generating saddle points in the merit function landscape of optical systems," Proceedings of SPIE, SPIE, 2005, p. 59620S-59620S-8.

[BORN99] M. Born and E. Wolf, Principles of Optics, Cambridge: Cambridge University Press, 1999.

[BRAU08] B. Braunecker, R. Hentschel, and H.J. Tiziani, Advanced Optics Using Aspherical Elements, Bellingham: SPIE Press, 2008.

[BUCH68] H.A. Buchdahl, Optical aberration coefficients, New York: Dover Publications, 1968.

[BURG09] J.H. Burge, Introductory Optomechanical Engineering - 9. Specifying Optical Components, 2009.

[CHAP98] H.N. Chapman and D.W. Sweeney, "A Rigorous Method for Compensation Selection and Alignment of Microlithographic Optical Systems," 23rd Annual International Symposium on Microlithography, 1998.

[OPTE07]   O. Consulting, Optische Technologien Wirtschaftliche Bedeutung in Deutschland, 2007.

[DGRO09]   J. DeGroote Nelson, R.N. Youngworth, and D.M. Aikens, "The cost of tolerancing," Proceedings of SPIE, vol. 7433, 2009, p. 74330E-74330E-12.

[DEWI05]   F. DeWitt IV and G. Nadorff, "Rigid body movements of optical elements due to opto-mechanical factors," Proceedings of SPIE, vol. 5867, 2005, p. 58670H-58670H-12.

[DILW08]   D.C. Dilworth, "New tools for the lens designer," Proceedings of SPIE, vol. 7060, 2008, p. 70600B-70600B-11.

[DIN00]   N.N., DIN ISO 10110, 2000.

[DOBS95]   S.J. Dobson and A. Cox, "Automatic desensitization of optical systems to manufacturing errors," Measuring Science Technology, vol. 1056, 1995.

[DOYL02]   K.B. Doyle, V.L. Genberg, and G.J. Michels, Integrated optomechanical analysis, SPIE Press, 2002.

[DYCK07]   H. Dyckhoff and T. Spengler, Produktionswirtschaft, Springer, 2007.

[FEDE62]   D.P. Feder, "Automatic Lens Design with a High-Speed Computer," Journal of the Optical Society of America, vol. 52, Feb. 1962, p. 177.

[FISC90]   R.E. Fischer, "Optimization of lens designer to manufacturer communications," Proceedings of SPIE, vol. 1354, 1990, p. 506.

[FISC08]   R.E. Fischer, B. Tadic-Galeb, and P.R. Yoder, Optical system design, McGraw-Hill, 2008.

[FORB08]   G.W. Forbes and C.P. Brophy, "Asphere, O asphere, how shall we describe thee?," Proceedings of SPIE, vol. 7100, 2008.

[FRAN11]   A. Frank, Connections in Combinatorial Optimization, New York: Oxford University Press, 2011.

[FRIE80]   B.R. Frieden, The computer in optical research, Springer, 1980.

[FUNC10]   M.C. Funck, J. Dolkemeyer, V. Morasch, and P. Loosen, "Design of a Miniaturized Solid State Laser for Automated Assembly," Proceedings of SPIE, 2010.

[GREY70] D.S. Grey, "Tolerance Sensitivity and Optimization," Applied Optics, vol. 9, Mar. 1970, p. 523.

[GROS05] H. Gross, Handbook of Optical Systems: Fundamentals of technical optics, Wiley-VCH, 2005.

[GROS07] H. Gross, H. Zügge, and M. Totzeck, Handbook of Optical Systems: Aberrations and correction of optical systems, John Wiley Professio, 2007.

[GÖRS99] D. Görsch, S. Kosub, and K.-P. Zocher, "Allgemeine Systeme der Toleranzgruppenoptimierung," Tagungsband zum 44. Internationalen Wissenschaftlichen Kolloquium (IWK' 99), vol. Band 1, 1999, pp. 411-417.

[HAFE03] H. Haferkorn, Optik:Physikalisch-Technische Grundlagen und Anwendungen, Weinheim: Wiley-VCH, 2003.

[HAFE84] H. Haferkorn, Synthese Optischer Systeme, Deutscher Verlag der Wissenschaften, 1984.

[HERI06] E. Hering and R. Martin, Photonik - Grundlagen, Technologie und Anwendung, Springer, 2006.

[HOPK66] H.H. Hopkins and H.J. Tiziani, "A theoretical and experimental study of lens centring errors and their influence on optical image quality," British Journal of Applied Physics, vol. 17, 1966, pp. 33-55.

[ISSH04] M. Isshiki, "Automated control of manufacturing sensitivity during optimization," Proceedings of SPIE, vol. 5249, 2004, pp. 343-352.

[ISSH07] M. Isshiki, D.C. Sinclair, and S. Kaneko, "Lens design: global optimization of both performance and tolerance sensitivity," Proceedings of SPIE, vol. 6342, 2007, p. 63420N-63420N-10.

[KARD05] O.J.W.F. Kardaun, Classical methods of statistics, Berlin Heidelberg: Springer Science & Business, 2005.

[KIDG01] M. Kidger, Fundamental optical design, SPIE Publications, 2001.

[KIDG04] M. Kidger, Intermediate optical design, SPIE Publications, 2004.

[KING78] R. Kingslake, Lens design fundamentals, Orlando: Academic Press, 1978.

[KING83] R. Kingslake, Optical System Design, Academic Press, 1983.

[KOCH78]   G. Koch, Donald, "A Statistical Approach to Lens Tolerancing," vol. 147, 1978, pp. 71-81.

[KREI10]   C.B. Kreischer, "The adversarial relationship between optical performance and scratch-dig," Design, vol. 7652, 2010, p. 76521K-76521K-8.

[KUPE93]   T.G. Kuper, "Global optimization for lens design: an emerging technology," Proceedings of SPIE, vol. 1781, 1993, pp. 14-28.

[LATY10]   S.M. Latyev, A.P. Smirnov, D. Frolov, A.G. Tabachkov, and R. Theska, "Providing target performance indices when automating the assembly of microscope," Journal of Optical Technology, vol. 77, 2010, pp. 38-41.

[LATY09]   S.M. Latyev, A.P. Smirnov, A.A. Voronin, B.S. Padun, and E.I. Yablochnikov, "The concept of an automatic assembly line for microscope objectives , based on," Design, 2009, pp. 436-439.

[LOOS11]   P. Loosen, R. Schmitt, C. Brecher, R. Müller, M. Funck, A. Gatej, V. Morasch, A. Pavim, and N. Pyschny, "Self-optimizing Assembly of Laser Systems," Production Engineering, 2011.

[LOTT06]   B. Lotter and H.-P. Wiendahl, Montage in der industriellen Produktion, Heidelberg: Springer, 2006.

[MARG99]   S. Magarill, "Optomechanical sensitivity and tolerancing," Proceedings of SPIE, vol. 3786, 1999, pp. 220-228.

[MAHA91]   V.N. Mahajan, Aberration theory made simple, Bellingham: SPIE Press, 1991.

[MCGU06]   J. McGuire James P., "Designing easily manufactured lenses using a global method," Proceedings of SPIE, vol. 6342, 2006, p. 634200-634200-11.

[KEHO10]   Michael Kehoe, "Tolerance Assignment for Minimizing Manufacturing Cost - OSA Technical Digest (CD)," International Optical Design Conference, Optical Society of America, 2010, p. ITuF5.

[MÄKI10]   J.-T. Mäkinen and S. Nollau, "Optics cost modelling and design optimization," Proceedings of SPIE, vol. 7717, 2010, pp. 1-15.

[MIL97]   N.N., MIL-PRF-13830B, 1997.

[OPTI10]   Optimax-Systems, "Optics Manufacturing Tolerances," 2010.

[PAHL06]   G. Pahl, W. Beitz, J. Feldhusen, and K.-H. Grote, Konstruktionslehre, Springer, 2006.

[PERI05]   J.-C. Perrin, "Making effective use of tolerancing," Proceedings of SPIE, 2005, p. 59620F-59620F-11.

[PINT80]   G. Pinto, "Nonstatistical method to evaluate tolerances of an optical system," Proceedings of SPIE, vol. 237, 1980, pp. 434-438.

[PLUM79]   Plummer, "Tolerancing for economies in mass production of optics," Proceedings of SPIE, vol. 181, 1979.

[RAY02]   S.F. Ray, Applied photographic optics: lenses and optical systems for photography, Oxford, Woburn: Focal Press, 2002.

[RAYL79]   F.R.S. Lord Rayleigh, "Investigations in Optics, with special reference to the Spectroscope," Philosophical Magazine, vol. 8(50), 1879, pp. 403-411.

[RIMM70]   M. Rimmer, "Analysis of Perturbed Lens Systems," Applied Optics, vol. 9, 1970, pp. 533-538.

[ROGE06]   J.R. Rogers, "Using global synthesis to find tolerance-insensitive design forms," Proceedings of SPIE, vol. 6342, 2006, p. 63420M-63420M-11.

[ROTH11]   F. Rothlauf, Design of Modern Heuristics, Berlin Heidelberg: Springer, 2011.

[SALE91]   B.E.A. Saleh and M.C. Teich, Fundamentals of Photonics, New York: John Wiley and Sons, 1991.

[SCHL96]   J. Schlichting, I. Balogh, H.J. Feige, L. Körner, U. Krüger, B. Schimmel, J.P. Schmidt, P. Stefan, and K. Winkler, "Production of high performance optics," Proceedings of SPIE, vol. 2774, 1996, pp. 623-630.

[SCHM10]   R. Schmitt and T. Pfeifer, Qualitätsmanagement, Carl Hanser Verlag, 2010.

[SCHU07]   G. Schuh, F. Klocke, C. Brecher, and R. Schmitt, Excellence in Production, Aachen: Apprimus-Verlag, 2007.

[SHAF80]   D. Shafer, "A modular method of Optical Design," J. Opt. Soc. America, vol. 70, 1980, p. 1608.

[SHAF89]   D. Shafer, "I plead the 5th," Proceedings of SPIE, vol. 1049, 1989.

[SHAF94]   D. Shafer, "Lens corrected to zero for all 3rd and 5th-order aberrations," 1994.

[SHAN97]   R. Shannon, The art and science of optical design, Cambridge University Press, 1997.

[SIEG02]   A. Siegel and G. Litfin, Deutsche Agenda Optische Technologien, Duisburg: WAZ-Druck, 2002.

[SING06]   J. Singh, Optical properties of condensed matter and applications, John Wiley and Sons, 2006.

[SMIT85]   W.J. Smith, "Fundamentals of establishing an optical tolerance budget," Proceedings of SPIE, vol. 531, 1985, pp. 359-366.

[SMIT92]   W.J. Smith, Modern Lens Design, McGraw-Hill, 1992.

[SMIT00]   W.J. Smith, Modern Optical Engineering, McGraw-Hill, 2000.

[SRIN07]   Srinivasan, Operations Research: Principles And Applications, New Delhi: PHI Learning Pvt. Ltd., 2007.

[STEP89]   D. Stephenson, "Forget About Finding the Problem ... Just Fix It," Proceedings of SPIE, vol. 1049, 1989, pp. 187-195.

[SZAP10]   S. Szapiel and C. Greenhalgh, "Lens solutions which increase manufacturing yield," Design, vol. 7652, 2010, p. 76521L-76521L-9.

[THOM80]   K.P. Thompson, "Aberration Fields in Tilted and Decentered Optical Systems," University of Arizona, 1980.

[THOR83]   E.K. Thorburn, "Tolerance and Techniques in High Precision Optical Assembly," Components, vol. 0406, 1983, pp. 113-118.

[TRIB64]   C. Tribastone, C. Gardner, and W.G. Peck, "Precision plastic optics applications from design to assembly," Proceedings of SPIE, vol. 2600, 1964, pp. 6-10.

[TRÄG07]   F. Träger, Handbook of lasers and optics, New York: Springer, 2007.

[TURN92]   T.S. Turner, "Vector aberration theory on a spreadsheet - analysis of tilted and decentered systems," Proceedings of SPIE, vol. 1752, 1992, pp. 184-195.

[VTUR09]   M. Van Turnhout, "A Systematic Analysis of the Optical Merit Function Landscape," Technische Universiteit Delft, 2009.

[VAND08]   R.J. Vanderbei, Linear programming: foundations and extensions, New York: Springer, 2008.

[WARN96]   H.J. Warnecke, Montage im flexiblen Produktionsbetrieb, Springer, 1996.

[WILL83]   R.R. Willey, "Economics in Optical Design, Analysis, and Production," Proceedings of SPIE, vol. 0399, 1983, pp. 371-377.

[WILL84]   R.R. Willey, "The Impact of Tight Tolerances and Other Factors on the Cost of Optical Components," Proceedings of SPIE, vol. 0518, 1984, pp. 106-111.

[WILL92]   R.R. Willey and M.E. Durham, "Maximizing production yield and performance in optical instruments through effective design and tolerancing," SPIE Critical Review, vol. CR43, 1992, pp. 76-108.

[WILL89]   D.M. Williamson, "Compensator selection in the tolerancing of a microlithographic lens," Proceedings of SPIE, vol. 1049, 1989, pp. 178-186.

[WILL10]   B. Willnauer, "Mass Production of Polymer Optics," Colloquium "Optics - Key Technology for the Future," 2010.

[YODE08]   P. Yoder, Mounting optics in optical instruments, SPIE Publications, 2008.

[YODE05]   P. Yoder, Optomechanical systems design, CRC Press, 2005.

[YOUN06]   R.N. Youngworth, "21st century optical tolerancing: a look at the past and improvements for the future," Proceedings of SPIE, vol. 6342, 2006.

[YOUN09]   R.N. Youngworth, "Tolerancing Forbes aspheres: advantages of an orthogonal basis," Proceedings of SPIE, vol. 7433, 2009, p. 74330H-74330H-12.

[YOUN01]   R.N. Youngworth and B.D. Stone, "Elements of Cost-Based Tolerancing," Optical Review, vol. 8, Jul. 2001, pp. 276-280.

[ZEMA10]   N.N., "ZEMAX ® Optical Design Program User's Manual," 2010.

[ZOCH85]   K.-P. Zocher, Adaptive und Selektive Montage in der flexiblen Fertigung, 1985.

III   Figures

IV   Abbreviations

A	$n*i$ (refractive index times marginal ray angle of incidence)
$\bar{A}$	$n*\bar{i}$ (refractive index times chief ray angle of incidence)
A	Boundary condition matrix
A_i	Assembly of components
AP	Assembly process
b	Boundaries of linear optimization
B_i	Sellmeier coefficients
C	Cost
C	Component
c	Curvature of a lens surface
$C_{1\text{-}3}$	Sellmeier coefficient C
C_i	Sellmeier coefficients
C_{nm}	Zernike coefficient
DLS	Damped Least Squares
f_i	Design function
F	Determining function
H	Lagrange invariant
h	Marginal ray height
i	Marginal ray angle of incidence
$\bar{i}$	Chief ray angle of incidence
i_{OAR}	Marginal ray angle of the optical axis ray
k	Symmetry factor

m	Number of modules
M	Selection matrix
M	Manufacturing process
MF	Merit function
n	Number of systems
n, n'	Refractive index in object and image space
n_C	Refractive index C-line
N_{comb}	Number of combinations for combinatorial assembly
n_D	Refractive index d-line
n_F	Refractive index F-line
$n_{Sellmeier}$	Refractive index approximation after Sellmeier
OPD	Optical Path Difference
OPD_{nom}	Nominal OPD
OPD_{spec}	OPD specification
OPD_{tol}	Tolerance allowance for OPD deviation
p	Price
PSF	Point Spread Function
P_{WC}	Worst case performance estimate
q	Performance
Q	Performance measure of a set of systems
Q^*	Performance deviation measure of a set of systems
q_{min}	Performance limit
q_{RSS}	Root sum square performance estimate
q_{WC}	Worst case performance estimate
R	Normal distance between perturbed surfaces

RSS	Root sum square
S_{eff}	Effective sensitivity
s_i	Seidel coefficients
S_i	Sensitivity
S_{I-IV}	Seidel sums
t	Time
t	Target of the merit function
t_{comp}	Time to compare numerical results
t_{eval}	Time to evaluate
t_i	Tolerance of parameter i
u	Abbe number
$u,\ u'$	Marginal ray angle before and after refraction
W	Wave aberration sum
W_{Aber}	Wavefront change due to aberrations
w_i	weight
W_{ijk}	Wave aberration coefficient
W_{Nom}	Nominal wavefront
W_{Tol}	Wavefront of disturbed system
x	Lens parameter
$\underline{x}$	Vector of lens parameters
X	Production parameter
y	Object height
y'	Image height
y_i	Marginal ray height at surface i
$\overline{y}_i$	Chief ray height at surface i

Z_n^m	Zernike polynomial
Δi	Tolerance limit of parameter i
δ_{OPL}	Optical path length difference
η	Normalized field size
θ	Azimuth angle
λ	Wavelength, Lagrangian multiplier
ρ	Normalized pupil size
σ	Vector aberration offset, standard deviation
φ	Tolerance distribution
ϕ	Probability
ϕ_{cum}	Cumulated probability
Γ	Success function

V   Appendix

1.1   Calculation of vector aberrations

The determination of the optical axis ray and in particular the angles of incidence i_{OAR} on individual surfaces is central to the calculation of vector aberrations. The calculation is based on determination of a generalized tilt parameter β describing the position of the surface.

The following equations can be used, where β_i is the tilt of surface i, c_i is the cuvature and δc_i is the decenter of surface i

1.1
$$\beta_{0,i} = \beta_i + c_i \cdot \delta c_i$$

The shifted pupil and image than iteratively calculate as

1.2
$$\delta q_i = \delta q_i' + \beta_{0,i} \cdot \Delta n_i \cdot y_i / H$$
$$\delta e_i = \delta e_i' - \beta_{0,i} \cdot \Delta n_i \cdot \overline{y}_i / H$$

With these given, the angle and position of the optical axis ray can be determined, before the angle of incidence of the optical axis ray on a surface can be calculated

1.3
$$u_{OARi} = \overline{u}_i \delta q_i + u_i \delta e_i$$
$$y_{OARi} = \overline{y}_i \delta q_i + y_i \delta e_i$$
$$i_{OARi} = u_{OARi} + y_{OARi} \cdot c_i - \beta_o$$

As the paraxial quantities remain constant for decentered or tilted elements W_{131} and the chief ray angle of incidence i do not change as well. Hence the effect of a single parameter perturbation can be expressed as

1.4
$$\frac{da_{131}}{dx_j} = \frac{w_{131,j}}{\overline{i}_j} \frac{\partial i_{OAR,j}}{\partial x_j} + \sum_{k \neq j} \frac{w_{131,k}}{\overline{i}_k} \frac{\partial i_{OAR,k}}{\partial x_j}$$

Conversely, the effect of parameter perturbations on a single surface can be expressed as

1.5
$$\frac{da_{131}}{dx_j} = \frac{w_{131,j}}{\overline{i}_j} \left\{ \frac{\partial i_{OAR,j}}{\partial x_j} + \sum_{k \neq j} \frac{\partial i_{OAR,k}}{\partial x_j} \right\}$$

The differentials of the OAR are easily calculated and can be combined into a statistical deviation such that

1.6

$$\frac{da_{131}}{dx_j} = \frac{w_{131,j}}{\bar{i}_j} \Delta i_{OAR,j} \quad , \quad \Delta i_{OAR,j} = \sqrt{\sum_k \frac{\partial i_{OAR,k}}{\partial x_j}}$$

may be regarded a measure of sensitivity. As w_{131} is a function of i the expression can be further simplified.

While for the first surface, the i_{OAR} is determined as the product of the generalized tilt parameter β and curvature, the following surfaces depend on the i_{OAR}'s as defined by preceding surfaces. Hence components and surfaces at the end of a lens have typically a larger influence as the resulting angle of the optical axis ray is larger. Using this during design optimization can be achieved by defining lens configurations with only one parameter disturbed to determine the angle of the optical axis ray.

In Zemax, the following Macro can be used to determine the quantities:

```
! The macro calculates third order vectorial aberrations for given
! decentrations of surfaces
! These surface de-centers are given as a vector delta in mm. Surface tilts
! can be translated
!

paraxmode = PMOD()
PARAXIAL ON
wl = PWAV()

ns = NSUR()
nsm = ns - 1

DECLARE y, DOUBLE,1,ns
DECLARE y_bar, DOUBLE,1,ns
DECLARE u, DOUBLE,1,ns
DECLARE u_bar, DOUBLE,1,ns
DECLARE i_bar, DOUBLE,1,ns
DECLARE dn,DOUBLE,1,ns
DECLARE Image_d,DOUBLE,1,ns
DECLARE Pupil_d,DOUBLE,1,ns
DECLARE beta,DOUBLE,1,ns
DECLARE delta,DOUBLE,1,ns
DECLARE u_OAR,DOUBLE,1,ns
DECLARE y_OAR,DOUBLE,1,ns
DECLARE i_OAR,DOUBLE,1,ns
DECLARE sigma,DOUBLE,1,ns
DECLARE a131,DOUBLE,1,ns
DECLARE a222,DOUBLE,1,ns
DECLARE A,DOUBLE,1,ns
DECLARE A_bar,DOUBLE,1,ns
DECLARE dun,DOUBLE,1,ns
DECLARE i_,DOUBLE,1,ns
DECLARE COMA,DOUBLE,1,ns
DECLARE ASTI,DOUBLE,1,ns


FOR i = 1,nsm,1
    delta(i) = 0
NEXT

!Define centrations
delta(4)=0.05
delta(8)=-0.1
```

```
delta(9)=delta(8)

!Calculate beta
FOR i = 1,nsm,1
    beta(i) = 0
    beta(i)=delta(i)*CURV(i)
NEXT

RAYTRACE 0, 0, 0, 1, wl
u0=RAYM(0) / RAYN(0)

FOR i = 1,ns,1
    y(i) = RAYY(i)
    u(i) = RAYM(i) / RAYN(i)

    IF (i>1)
    i_(i)= y(i)*CURV(i)+u(i-1)
    ELSE
    i_(i)= y(i)*CURV(i)+u0
    ENDIF

    A(i) = INDX(i-1)*i_(i)
    dn(i) = INDX(i)-INDX(i-1)
    dun(i) = u(i)/INDX(i) - (RAYM(i-1)/RAYN(i-1)/INDX(i-1))
NEXT
    !PRINT "y      u      i    ybar    ubar    ibar"


RAYTRACE 0, 1, 0, 0, wl
u_bar0 = RAYM(0) / RAYN(0)

FOR i = 1,ns,1
    y_bar(i) = RAYY(i)
    u_bar(i) = RAYM(i) / RAYN(i)

    IF (i>1)
    i_bar(i) = y_bar(i)*CURV(i)+u_bar(i-1)
    ELSE
    i_bar(i) = y_bar(i)*CURV(i)+u_bar0
    ENDIF

    A_bar(i) = INDX(i-1)*i_bar(i)
    !PRINT y(i)," ",u(i), " ",i_(i),"   ",y_bar(i)," ",u_bar(i), " ",i_bar(i)
NEXT

pyc = RAYY(1)
puc = RAYM(1) / RAYN(1)
Lagrange = INDX(1) * (y(1) * puc - u(1) * pyc)
    PRINT Lagrange

    PRINT "delta n      d(u/n)     COMA"

FOR i = 1, ns, 1
    COMA(i) = -A(i)*A_bar(i)*dun(i)*y(i)/WAVL(wl)/2*1000
    ASTI(i) = -A_bar(i)*A_bar(i)*dun(i)*y(i)/WAVL(wl)/2*1000
    PRINT dn(i)," ",dun(i)," ",COMA(i)
NEXT

Image_d(1)=0
Pupil_d(1)=0

FOR i = 2, ns, 1
    Image_d(i) = Image_d(i-1) + y(i)*dn(i)*beta(i)/Lagrange
    Pupil_d(i) = Pupil_d(i-1) - y_bar(i)*dn(i)*beta(i)/Lagrange
NEXT


FOR i = 1, ns, 1
```

```
u_OAR(i) = Image_d(i)*u_bar(i) + Pupil_d(i)*u(i)
y_OAR(i) = Image_d(i)*y_bar(i) + Pupil_d(i)*y(i)

!Falls i = 1 muss statt u_OAR(0) u(0)=u0 eingesetzt werden

IF (i>1)
i_OAR(i) = u_OAR(i-1)+CURV(i)*y_OAR(i)-beta(i)
ELSE
i_OAR(i) = u0+CURV(i)*y_OAR(i)-beta(i)
ENDIF

sigma(i) = -i_OAR(i)/i_bar(i)
a131(i) = COMA(i)*sigma(i)
a222(i) = ASTI(i)*sigma(i)
NEXT

PRINT "        surf        y_bar(i)        u_bar(i)        i_bar(i)        u_OAR(i)
y_OAR(i)        i_OAR(i)        sigma(i)"

FOR i = 1, ns, 1
    FORMAT 12.5
    PRINT i," ",y_bar(i),"   ",u_bar(i), "   ",i_bar(i), "   ",u_OAR(i),"
",y_OAR(i),"   ",i_OAR(i),"   ",sigma(i),"   ",a131(i)
NEXT

V_COMA=0
FOR i= 1, ns, 1
    V_COMA = a131(i)+V_COMA
    V_ASTI = a222(i)+V_ASTI
NEXT

PRINT "Image displacement: ",Image_d(ns)*y_bar(ns)
PRINT "Total Coma: ",V_COMA
PRINT "Total Asti: ",V_ASTI
```

1.2   Tolerance laws for uniform distribution

In chapter 4 cumulated probability curves are given for Gaussian tolerance distributions of the matching parameters. The curves for uniform distributions look different and are depicted here.

Figure 82
Scaling of perfectly
matched tolerance
regions

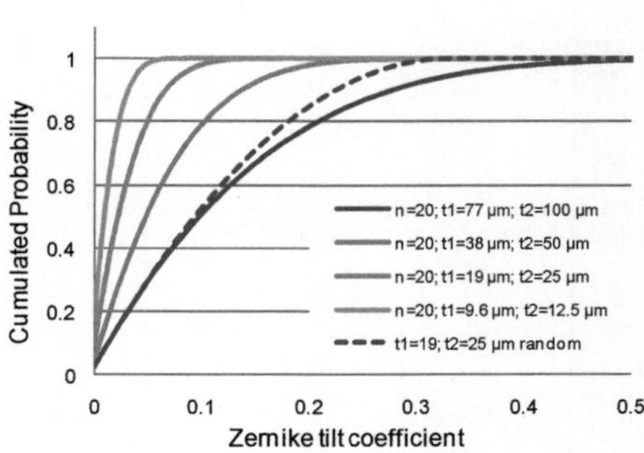

If tolerances do not match, the following curves result for a linear performance function:

Figure 83
Scaling of imperfectly matched tolerance regions

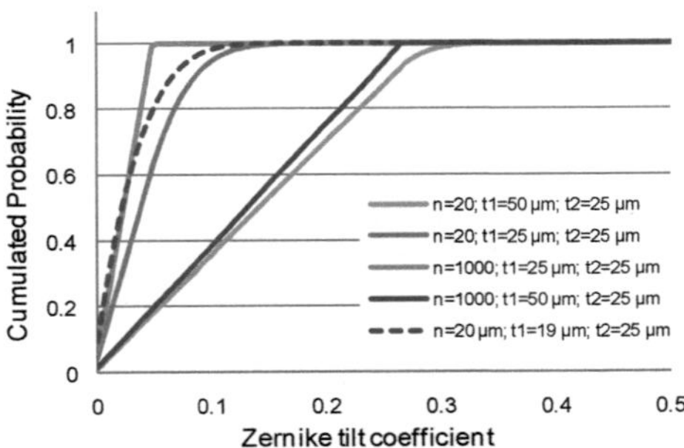

Scaling imperfectly matched tolerances but keeping the tolerances in their proportion results in the following graphs.

Figure 84
Scaling of imperfectly matched tolerance regions

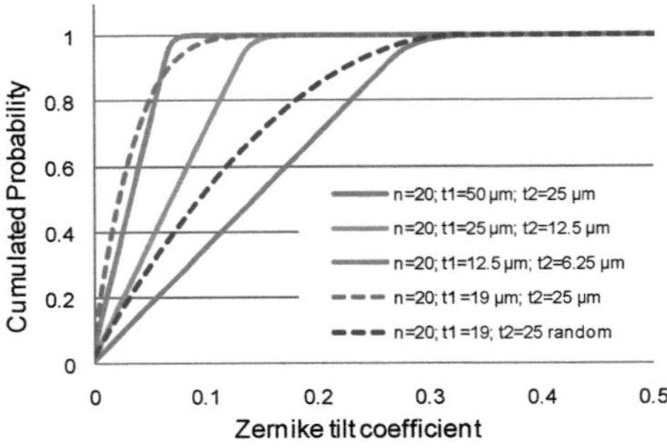

The curves show, that in this case, a proportional reduction of the tolerances also reduces the observed error in linear proportion.

1.3   Lens data of the global optimization

For the global optimization example lenses were designed to have 8 mm entrance pupil diameter at 1° field and 632 nm wavelength. Restrictions were made with respect to thicknesses.

The results of the optimized systems are as follows:

System 1

Figure 85
Lens data of
system 1

Surface	Radius	Curvature	Thickness	Glass
0	0.00000	0.00000	Infinity	
1	0.00000	0.00000	14.99952	
2	103.93145	0.00962	4.99981	SILICA
3	-17.19075	-0.05817	3.14344	
4	-10.16792	-0.09835	4.08827	SILICA
5	-18.78937	-0.05322	3.36356	
6	17.83114	0.05608	5.00013	SILICA
7	-23.23889	-0.04303	15.00030	
8	0.00000	0.00000	0.00000	

System 2

Figure 86
Lens data of
system 2

Surface	Radius	Curvature	Thickness	Glass
0	0.00000	0.00000	Infinity	
1	0.00000	0.00000	14.99980	
2	36.56921	0.02735	5.00009	SILICA
3	-148.19243	-0.00675	2.33223	
4	18.42819	0.05426	4.38770	SILICA
5	58.60538	0.01706	0.99988	
6	10.37085	0.09642	3.75653	SILICA
7	17.11093	0.05844	12.03737	
8	0.00000	0.00000	0.00000	

System 3

Surface	Radius	Curvature	Thickness	Glass
0	0.00000	0.00000	Infinity	
1	0.00000	0.00000	15.00008	
2	10.67724	0.09366	4.54443	SILICA
3	7.70996	0.12970	0.99586	
4	15.21127	0.06574	3.87510	SILICA
5	-28.35885	-0.03526	1.00000	
6	6.32364	0.15814	3.21398	SILICA
7	7.10691	0.14071	11.99964	
8	0.00000	0.00000	0.00000	

System 4

Surface	Radius	Curvature	Thickness	Glass
0	0.00000	0.00000	Infinity	
1	0.00000	0.00000	14.99998	
2	20.00000	0.05000	3.00000	SILICA
3	10.66987	0.09372	15.00098	
4	39.45860	0.02534	4.00000	SILICA
5	-26.59964	-0.03759	1.00121	
6	23.26391	0.04299	4.00000	SILICA
7	-57.63321	-0.01735	22.02959	
8	0.00000	0.00000	0.00000	

System 5

Surface	Radius	Curvature	Thickness	Glass
0	0.00000	0.00000	Infinity	
1	0.00000	0.00000	9.99999	
2	-8.67766	-0.11524	2.29574	SILICA
3	-10.50187	-0.09522	0.99931	
4	48.65607	0.02055	3.49694	SILICA
5	-25.22366	-0.03965	8.53984	
6	13.89218	0.07198	4.40540	SILICA
7	114.00503	0.00877	14.20196	
8	0.00000	0.00000	0.00000	

1.4   Details on the MicroSlab example

In order to conduct combinatorial assembly, the deviations of components were measured in advance. The critical parameters are height of the emission of the laser diode as well as centration of lenses A and B. For both measurements, digital image processing is employed.

The following figure shows images of the laser diode and lenses that were acquired during this process.

Figure 90
Images of laser
diode and lens

In order to determine the emission height and the centration of the lenses, the acquired images were processed and edge detection used to recognize features of the components. For the lenses, the entire contour in a side-cut is imaged (both sides) and then the centration calculated.

Figure 91 shows histograms of centering errors for both types of lenses derived from measurements of 50 lenses each.

Figure 91
Centering errors of
lenses A and B with
a nominal value of
1.5 mm

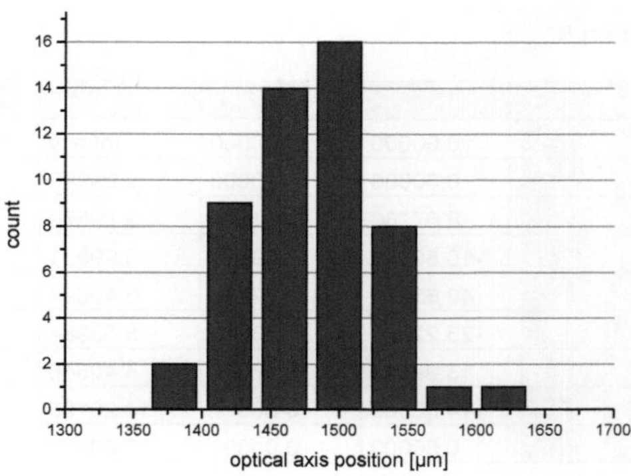

An additional application of combinatorial assembly to the MicroSlab laser ist he mathcing of micro lens arrays and focusing lenses. Because the width of the pump line only depends on the divergence of the lens array and the focal length of the lens alignment is physically impossible. The distance between these two elements as well as any other movements have no influence on the width of the pump line. The width b is given as

0.7 $$b = \alpha \cdot f$$

Figure 92 shows the top view of the pump optics of MicroSlab illustrating the working principle of the micro lens array homogenization.

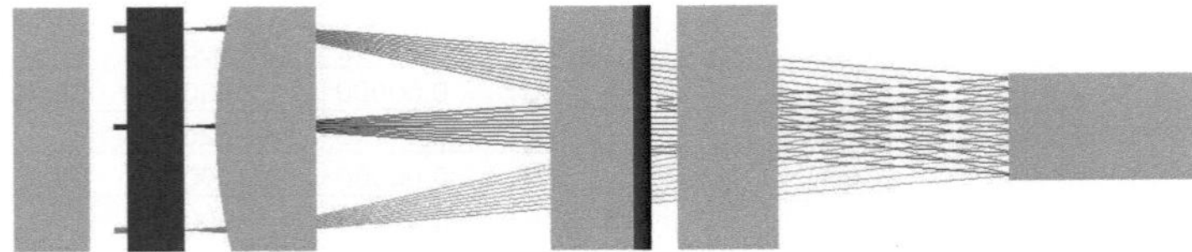

Figure 92
MicroSlab pump optics top view showing homogenization

Adjusting the width of the pump line to the laser crystal can help to increase the efficiency and reduce thermal effects due to bad overlap of crystal and beam.

Accurately measuring the divergence of the lens arrays is however quite difficult and as the effect of misaligned beam and crystal did not pose any problems during operation combinatorial assembly was not employed.

1.5   Details on the laser focusing lens example

The lens data of the laser focusing lens:

Figure 93
Lens data of the
laser focusing lens

Surface	Radius	Curvature	Thickness	Glass	Conic
1	0	0.00000	47.03337		
2	0	0.00000	6.00000	SILICA	
3	-45.9	-0.02179	0.35000		
4	0	0.00000	4.90000	SILICA	
5	-53.5	-0.01869	0.31000		
6	0	0.00000	2.00000	SILICA	
7	48.8	0.02049	4.83000		
8	0	0.00000	6.00000	SILICA	
9	-45.9	-0.02179	20.00000		
10	0	0.00000	2.00000	SILICA	
11	23.775	0.04206	41.62502		
12	0	0.00000	12.00000	SILICA	
13	-48.3475	-0.02068	1.00000		
14	0	0.00000	8.50000	SILICA	
15	-72.0795	-0.01387	7.00000		
16	32.62	0.03066	15.40000	S-LAH64	-1.71664
17	0	0.00000	24.61711		

Tolerances used to compute the sensitivities and change table:

Figure 94
Opto-mechanical
tolerances of the
laser focusing lens

Lens tolerances		
CURV	12	0.00001
TTHI	15	0.05
CURV	15	0.000083332
TTHI	18	0.05
CURV	18	0.00001
TTHI	21	0.05
CURV	21	-4.13668E-05
TTHI	24	0.05
CURV	24	0.00001
TTHI	27	0.05
CURV	27	-2.77472E-05
TTHI	30	0.04
CURV	30	3.88672E-05
CURV	33	0.00001

Assembly tolerances		
TPAR2	11	0.0084
TPAR3	11	0.01
TPAR2	16	0.01
TPAR3	16	0.01
TPAR2	21	0.01
TPAR3	21	0.01
TPAR2	26	0.01
TPAR3	26	0.01